Lecture Notes in Intelligent Transportation and Infrastructure

Series Editor

Janusz Kacprzyk, Systems Research Institute, Polish Academy of Sciences, Warsaw, Poland

The series "Lecture Notes in Intelligent Transportation and Infrastructure" (LNITI) publishes new developments and advances in the various areas of intelligent transportation and infrastructure. The intent is to cover the theory, applications, and perspectives on the state-of-the-art and future developments relevant to topics such as intelligent transportation systems, smart mobility, urban logistics, smart grids, critical infrastructure, smart architecture, smart citizens, intelligent governance, smart architecture and construction design, as well as green and sustainable urban structures. The series contains monographs, conference proceedings, edited volumes, lecture notes and textbooks. Of particular value to both the contributors and the readership are the short publication timeframe and the world-wide distribution, which enable wide and rapid dissemination of high-quality research output.

More information about this series at http://www.springer.com/series/15991

Dmytro Gulevets · Artur Zaporozhets ·
Volodymyr Isaienko · Kateryna Babikova

Environmental Control for Ensuring Cities Safety

Springer

Dmytro Gulevets
National Aviation University
Kyiv, Ukraine

Volodymyr Isaienko
National Aviation University
Kyiv, Ukraine

Artur Zaporozhets
Institute of Engineering Thermophysics
National Academy of Sciences of Ukraine
Kyiv, Ukraine

Kateryna Babikova
National Aviation University
Kyiv, Ukraine

ISSN 2523-3440 ISSN 2523-3459 (electronic)
Lecture Notes in Intelligent Transportation and Infrastructure
ISBN 978-3-030-66709-2 ISBN 978-3-030-66710-8 (eBook)
https://doi.org/10.1007/978-3-030-66710-8

This Springer imprint is published by the registered company Springer Nature Switzerland AG
The registered company address is: Gewerbestrasse 11, 6330 Cham, Switzerland

Abbreviations

AM	Arithmetic mean
API	Air pollution index
APIn	Application programming interface
CGO	Central Geophysical Observatory
DB	Database
DF	Degrees of freedom
EMR	Electromagnetic radiation
ERS	Earth remote sensing
ES	Environmental safety
GIS	Geographic information system
IA	International airport
MHU	Ministry of Health of Ukraine
MPC	Maximum permissible concentration
MPL	Maximum permissible level
MS	Mean sum squares
NMP	Noise monitoring post
PMP	Pollution monitoring post
RS	Runway strip
SBS	State building standard
SESU	State Emergency Service of Ukraine
SLAA	Surface layer of atmospheric air
SS	Sum squares
SSS	State sanitary standard
UA	Urban area
V	Variable

Chapter 1
Research of Scientific Bases and Methodologies for Evaluating the State of Ecological Safety in Urban Areas

Abstract Ensuring the ecological safety of an urbanized area, namely the development of special measures to maintain the quality of the urban environment at the required level, is a relatively new area of research. The substantive content of the scientific and conceptual basis for the ecological safety of an urbanized territory is considered on the example of the state of quality of the surface layer of atmospheric air, taking into account individual factors of chemical and physical pollution. Determination of the quality of the surface layer of atmospheric air includes the following stages: an inventory of factors and sources of impact; collection and generalization of information on the levels of exposure to adverse factors on the population and the environment; determination of points or zones within the residential area, for which it is necessary to reduce man-made impact.

Keywords Urbanized area · Surface layer of atmospheric air · Ecological safety · Technogenic impacts

1.1 Modern Concept of UA

Currently, the term "urbanization" is used in several different meanings [1–3], the main of which are:

- growth and development of cities;
- increase in the proportion of urban population in a country, region, world;
- acquisition by the countryside of external and social features characteristic of the city;
- process of increasing the role of cities in the development of society.

Urbanization is a consequence of the scientific and technological revolution, one of the most important prerequisites for further development. The expansion of the urbanization process, which is not controlled, throughout the territory of individual countries and large regions leads to disruption of the normal activity of the planet's biogeocenotic cover [4–6]. At UA it is necessary to ensure the conservation and

D. Gulevets et al., *Environmental Control for Ensuring Cities Safety*,
Lecture Notes in Intelligent Transportation and Infrastructure,
https://doi.org/10.1007/978-3-030-66710-8_1

quantity of natural resources capable of reproducing not only biomass as food raw materials, but also clean air, water, fauna, and satisfy human needs.

Urbanization can be considered as one of the forms of human adaptation to the environments [7]. The growth of cities, the development of industry and vehicles into them is the objective reality of the modern world.

Under the "urbanized territory" refers to the ratio of UA to the total area of the region (territory, region, district).

For UA, the following parameters are usually calculated:

- population density in UAs,
- provision of citizens with green spaces and landscaping of UAs,
- development of transport infrastructure,
- provision of water and recreational resources.

In addition, an integrated environmental assessment has been introduced for UA, which is determined by the total assessment of the degree of influence of technogenic factors of the stability of natural complexes and components [8–10].

Specific features of UA are:

- high level of population density, which excludes the conservation of natural ecosystems in a given territory;
- formation of a new artificially supported system (the so-called urban system) with certain structural and functional organization;
- continuous monitoring of the quality of the urban environment.

In this regard, the structural and functional organization of cities is greatly complicated and there is an urgent need to develop scientifically sound control methods (environmental monitoring) that can become the basis for effective management of the city's environmental state (environmental management).

The nature of the functioning of the urban system is determined by the dynamics of the processes in the subsystems, the intensity of direct and reverse, negative and positive relationships between them. So, the natural system is characterized by geochemical and biological processes in the city—the transformation of rocks, participation in the regeneration of oxygen and water due to photosynthesis, etc. The anthropogenic subsystem is characterized by the creation and further development of structures of technogenic origin, the transformation of the natural landscape into a cultural one, etc. [11, 12].

The main difference between the effects of natural and anthropogenic systems is the ability of the natural subsystem of UA to self-regulation. The anthropogenic system, on the contrary, is completely dependent on the natural system. Moreover, a human, as an element that simultaneously belongs to both systems, creates an anthropogenic system, greatly changing the natural one, which deprives him of the ability to self-regulation.

Anthropogenic transformation of the environment, as well as manifestations of the toxicity of this environment for living objects, including for the human body, have reached maximum expression in places of concentration of production processes and high population density, especially in large industrial centers and megacities [13–15].

Cities are the center of major industrial pollutants that affect the environment. In most urban settlements, excesses of permissible concentrations of harmful substances in the atmosphere, water, and soil are recorded, and therefore from 10% of UAs (according to optimistic estimates) to 15% deserve to be recognized as an environmental emergency zone [9, 16–18].

Another aspect of the modern concept of the urban system is the development of special measures to maintain the quality of the urban environment at the required level.

A feature of the development of UA is the growing contradiction between the growth of the urban population and the desire of people to more comfortable living conditions in cities. The environmental conditions of the twentieth century, which were perceived as comfortable, are currently considered unacceptable. In this regard, there is a constant need for the development of new, science-based methods for improving the ecology of the city.

Based on the analysis of trends in the environmental situation in large cities, there are three stages of events aimed at improving the quality of the urban environment [19].

The first stage is associated with the widespread introduction of treatment facilities and improving the technical level of their maintenance.

The second stage is mainly based on the introduction of a comprehensive integrated recycling of all types of waste from large cities.

The third stage is associated with the widespread introduction of non-waste and low-waste technologies.

With a huge concentration of industry (in cities with more than 1 million citizens), it is impossible to achieve the necessary level of emissions and discharges treatment even with high technical efficiency.

The role of human in the city is a strong deformation of the output ecosystems and the creation of new specific elements (development, communications, etc.). For example, the influence of the city on the physicomechanical properties of the rocks of the upper horizon of the lithosphere can be traced to a depth of 20–50 m [20–22]. Most actively rocks change under the influence of static and dynamic loads, watering and drainage, vibration and electromagnetic radiation. Urban development affects groundwater to a depth of 100–150 m [23, 24]. Their standing level, temperature and chemical composition are changing, the equilibrium interaction of surface and groundwater is disturbed.

Regulation of river flow in cities leads to a decrease in their ability to cleanse themselves. It is estimated that a city with 1 million citizens consumes 400–500 million m^3 of water per year. To regenerate such a quantity of water, a catchment area of at least 20 thousand km^2 is required (with a surface runoff module of 0.2 l from 1 km/day).

In the process of production activities, the city changes the quality of the atmosphere, causing a change in microclimatic conditions. The nature of environmental events in cities depends, in addition, on solar radiation, determines the rate of self-cleaning of the environment, microclimatic conditions and the associated processes of formation of temperature inversions, clouds and smog.

One of the effective means of improving the environment of the city is landscaping. The role of green spaces in reducing the negative impact of the environment lies in their ability to reduce the adverse effects for humans of man-caused factors. Plants clean, moisturize, and enrich the air in cities, reduce the strength of wind and noise, and change the radiation and temperature conditions [25–27].

On average, 1 ha of plantations in one hour absorbs 8 kg of carbon dioxide CO_2, that is, the amount that 200 people exhale. Green spaces significantly reduce the concentration of harmful gases recognized as dangerous air pollutants such as sulfur dioxide SO_2 and hydrogen sulfide H_2S [25].

Groups of trees retain 21–86% of dust, and reduce air pollution by harmful microorganisms by 19–44%. It is estimated that vegetation absorbs from air and binds 50–60% of toxic gases, while: atmospheric moisture—5–20%, soil—5–10%, water bodies and animals—less than 5% [28–32].

At determining the strategy and tactics for improving UA, many researchers put the strengthening of urban ecological systems in the first place, for which they recommend increasing the area of green spaces in cities. At the same time, a peculiar paradox arises: it is in cities where plants with their sanitary-hygienic and recreational functions are most necessary that the greatest obstacles to their normal life are observed, consisting in a reduction in their life expectancy, reduced productivity, damage to needles and leaves.

1.2 Identification of the Main Pollution Factors of an UA

Each UA in many respects is a completely unique natural ecological system. In this regard, it is necessary to summarize numerous data on the main environmental factors.

Urbanization is always accompanied by the introduction of a large number of foreign components into the environment, that is, its intense pollution.

At the same time, the level of pollution determines the comfort of human life UA. Thus, on the one hand, pollution of the urban environment cannot be avoided, on the other hand, it must be kept at a certain level, which allows quite favorable conditions for human life. Otherwise, pollution fundamentally changes the functional characteristics of the urbanized system.

UA pollution is understood as the changes introduced into nature by human activities affecting the organic world. There are direct, indirect, positive and negative effects. Direct effects is directed on living organisms; indirect effects aimed at climate change, the physical and chemical state of the atmosphere, water bodies, the structure of the surface of the earth, flora and fauna; a negative effect is directed at the oppression or extinction of organisms; positive effect is directed on creating favorable conditions for the development of other organisms.

Based on the definitions given in the environmental dictionaries, pollution can be considered almost all elements of the urbanized system, including those necessary to ensure human life. The amount of pollutants to be controlled is constantly changing. The main pollutants include dust, carbon monoxide (CO), nitrogen oxides (NO_2) and sulfur (SO_2), hydrogen sulfide (H_2S), fluorine compounds (F) and inorganic hydrocarbons.

In recent years, interest in trace elements and their compounds [30, 33] and pesticides has sharply increased. This is due both to the facts of the manifestation of acute toxic effects, and to the negative effect of small concentrations of these substances on organisms. The weak ability of the urbanized system to self-clean from pesticides and a number of metals is significant [30]. This applies to the migration of highly toxic elements in ecological systems assigned to the 1st and 2nd hazard classes: Arsen (As), Cadmium (Cd), Cobalt (Co), Chrome (Cr), Copper (Cu), Mercury (Hg), Nickel (Ni), Plumbum (Pb), Zinc (Zn), etc. The content of these elements characterizes their mobility in the soil environment and is an indicator of their migration in ecological systems.

Technogenic geochemical anomalies are formed in various environments and, according to this criterion, they can be divided into lithochemical, hydrochemical, atmogeochemical, and biogeochemical. The latter can be subdivided into phyto-, zoo- and anthropogeochemical anomalies. Since pollutants are partially included in the biological cycle of substances, technogenic anomalies usually cover the sphere of influence of several media and form technogenic ranges and dispersion fluxes that are complex in shape, length, composition and nature of the differentiation of substances.

The stage of identification of the main environmental pollution factors of UA includes the following processes: inventory of factors and sources of exposure; collection and synthesis of information on the levels of exposure to adverse factors; identification of points or zones within the residential area for which it is necessary to reduce the industrial impact. The identification of the main environmental factors in the environment can be carried out using various methods: monitoring and diagnostics of environmental quality, modeling the behavior of chemical and physical factors (Fig. 1.1).

The highest degree of chemical load on the environment in cities is associated with dust and gas emissions into the atmosphere. Atmospheric pollutants by origin can be primary (waste products of enterprises, automobile engines) and secondary, formed in the free atmosphere as a result of chemical, photochemical, physico-chemical reactions between pollutants and atmospheric components [34–36].

In the conditions of the city, the geochemical identity of individual industries is smoothed out due to the imposition of individual anomalies as a result of the compact location of industrial enterprises.

Along with industry and other sources of pollution (energy and heating systems, waste incineration), vehicles have a very large share in urban air pollution [37].

The danger of road transport in air pollution does not only consist of the absolute mass of emissions. Great importance has fact that the exhaust gases enter directly

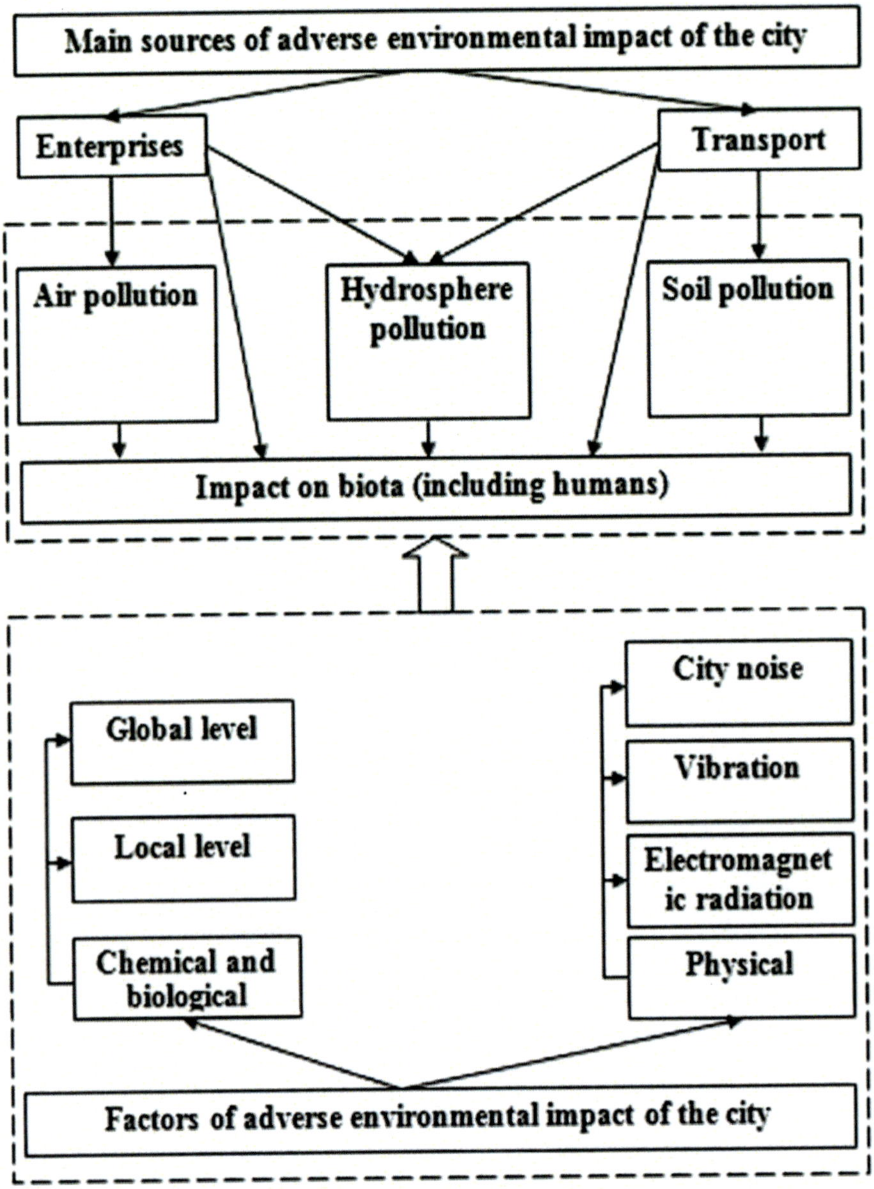

Fig. 1.1 Main sources and factors of technological impact on the environment of the urbanized territory

into the atmosphere, where the wind speed is low and the gases are poorly dispersed, in contrast to the emissions of enterprises due to high factory pipes [38–40].

In addition, heavy gases make up the bulk of the exhaust ingredients, prevents their dispersion, especially in places of calm wind. It is established that the quantitative and qualitative composition of car exhausts depends on the type of engines of the cars themselves. The most significant environmental damage is lead. It is established that during the combustion of fuel in internal combustion engines about 75% of the lead contained in gasoline is carried out with exhaust gases.

Thus, with a content of 1 g of plumbum in 2 L of gasoline and with a traffic density of 1000 cars per hour, 30–40 g of plumbum per hour is emitted at each kilometer of the road. Plumbum compounds settle mainly on soil and vegetation, and then are included in the food chain.

A high level of concentrations of chemical elements in precipitation is correlated with their significant content in atmospheric air, which in turn causes pollution of other media (soil, water), creating conditions for biological accumulation. Thus, the influence of the whole complex of pollution sources on the urban environment leads to the formation of geochemical anomalies, expressed as interconnected zones of increased concentrations of chemical elements in all environments: atmospheric air, water, soil, bottom sediments, living organisms. Due to selective absorption, geochemical anomalies in plants do not always reflect the true geochemical picture.

1.3 Modern Methods for Assessing the Quality of the Urban Environment

The lack of a sufficiently developed system of objective indicators of the rational use of territorial resources is carried out, as a rule, by increasing the area of the city with the development of mainly new territories, often agricultural land.

The most important criterion for the efficient use of UAs is considered the cost of their development and indicators of unit costs for the placement of construction objects.

In the practice of city design, the use of specific indicators of territorial expenditures is most often applied to residential territories, and not to the city as a whole.

In recent years, in order to intensify the use of UA, a number of regulatory measures have been developed to increase the density of its development by increasing the number of storeys of residential buildings, but this gives no more than 10% savings in the growth of the territory that is used. The planning methods aimed at increasing the intensity of development of UAs—functional ordering, reconstruction, enlargement of the planning module, the use of special methods of development, etc., acquired development.

As a result of the accumulation of project experience and the expansion of theoretical ideas about the city, it becomes more and more obvious that the economic criterion, with the help of which it would be possible to evaluate the quality of urban space, does not ensure the achievement of a comprehensive result—an increase in the ES of UA as a functionally integrated system that is developing.

The most important direction in achieving the efficient use of territories by functional planning means, especially in large, large and largest cities, was the increase of ES, including the optimization of the citywide system of functional zoning, monitoring the state of ES, production zones, functionally unused territories [41–43].

Great opportunities lie in the increase in the efficiency of use of territories in the underground urbanization of industrial production zones, which provides for the creation of underground and buried rooms, warehouses, garages, etc., at laying underground utilities, building underground engineering structures, organizing underground transport systems for passengers cargo [44, 45].

On the other hand, it should be noted that the intensification of the use of urban space, the reduction of functional relationships leads to a deterioration in the state of ES.

To a certain extent, this process can be regulated by traditional urban planning means—the removal of individual industrial enterprises polluting the environment, differentiating traffic flows, landscaping, methods of organizing space inside the main functional elements, etc.

The structure of the criteria for an effective functional planning organization of the city plan is based on two main approaches—normative and optimization.

The structure is as follows:

- intensity of use of local territories. The criterion is based on a system of indicators (building coefficient, population density, density of fixed assets, local density of fixed assets, utilization of underground space (planning and cost), average number of storeys, characteristic of the state of fixed assets—physical, functional (moral) and external (economic) depreciation and dynamics of profitability) [46, 47];
- efficiency of using the territory of the city. This criterion is based on the optimization of the following indicators: minimizing the average communication radius of UA, minimizing the time spent on moving, optimizing the linear density of settlement, minimizing energy costs, spatial compatibility, polyfunctioning of the using of the territories of the city plan [48].

Chemical pollution. Air quality standards determine the permissible limits for the content of harmful substances, both in the production (intended for the placement of industrial enterprises, pilot production research institutes, etc.), and in the residential zone (intended for the placement of housing stock, public buildings and structures) inhabited points.

The basis for regulating the atmospheric air quality of settlements is hygienic standards—MPC of atmospheric pollution by chemical and biological substances, the observance of which ensures the absence of their direct or indirect impact on the health of the population and living conditions. Each object, which is a source of atmospheric air pollution, must provide a control and monitoring system for atmospheric air pollution in the residential area in the zone of emissions of this object. The control and monitoring system must comply with the requirements [30, 33].

The placement of observation posts, the list of pollutants which must be controlled, methods for their determination, the frequency of sampling of atmospheric air are agreed with the bodies and institutions of state sanitary supervision.

The results of departmental control and monitoring of atmospheric air pollution are presented to the objects and institutions of state sanitary supervision in an agreed form. With the joint presence in the atmospheric air of several substances with a cumulative effect, the sum of their concentrations should not exceed 1 at calculating by the formula (1.1):

$$\frac{C_1}{MPC_1} + \frac{C_2}{MPC_2} + \cdots + \frac{C_n}{MPC_n} \leq 1, \tag{1.1}$$

where C_1, C_2, ..., C_n are the actual concentrations of substances in atmospheric air, MPC_1, MPC_2, ..., MPC_n are the maximum allowable concentrations of substances.

It was shown in [33] that during assessing the total air pollution, the hazard class of pollutants, the multiplicity of exceeding the MPC of the impurity, the nature of the combined effect of substances that are jointly present in the air should be taken into account.

In hygienic research, a more informative indicator—API of cities, calculated by formula (1.2):

$$API = a_i \left(\frac{b_i \cdot C_i}{MPC_i - P_i} \right), \quad i = \overline{1, N} \tag{1.2}$$

where a_i—coefficient taking into account the hazard class of the i-th substance; b_i—coefficient taking into account the effect of the action and the first substance; P_i—value that takes into account the threshold of action of the first substance, N—the totality of pollutants.

The basic unit of measurement is the linear deviation of the concentration of substances from their action thresholds. In the formula (1.2), the difference between the concentration of the substance and the threshold of its action characterizes the linear deviation of the pollution level from the sanitary norm, regardless of the direction of the deviation. The calculation formula (1.2) in terms of its mathematical content allows the possibility of a negative result (or zero). The first is possible when the concentration of all impurities is below their thresholds. Such a result can be a sign of the absence of pollution hazard, but to preserve the correctness of computational operations, the introduction of an additional condition that allows screening of impurities is required. With such a record, a zero result is possible in the case of mutual compensation of positive and negative deviations, which contradict the principle of the pollution assessment criterion. It should be noted that in the considered indicator only a part of the types of the effect of the general action of substances is taken into account—summation, synergy. In addition, formula (1.2) is constructed so that the threshold value is limited only by values of zero or MPC, which artificially narrows the scope of its application.

The application of the API formula (1.2) makes sense only if the condition is fulfilled—the concentration of all impurities should be higher than their MPC. In fact, the most typical case is when from of the aggregate (N) of pollutants, the pollution level for some of them (N_1) exceeds the standard $k > 1$, $k = C/MPC$, and for others ($N_2 = N - N_1$) it is lower ($k < 1$). In addition, with the general presence of harmful substances in the air, their combined effect is manifested in the effects of summation, potentiation, antagonistic or independent action.

At concentrations close to MPC, taking into account the hazard class of substances and the effect of their total exposure, a change in the ratio $k < 1$ to the ratio $k > 1$ and vice versa is possible. Therefore, ignoring the hazard class of substances at concentrations below the MPC and the effect of the general action of substances seems incorrect.

The practical need for a comparative assessment of the dynamics of changes in the quality of the atmosphere led to the development of indicators of its physical condition characterizing the presence, composition and quantitative content of harmful impurities in the air. One of the API models is widely used in the national system of monitoring and control of atmospheric pollution (formula (1.3)):

$$API = \sum \left(\frac{C_i}{MPC_i}\right)^{m_i}, \tag{1.3}$$

where m_i is an exponent taking into account the hazard class of the impurity.

In the future, this model was modernized. An attempt was made to take into account the effect of summation, synergism and removal of the restriction on a fixed amount of impurities that are taken into account:

$$API = \sum_{i=1}^{N_1}\left[\left(\frac{C_i}{MPC_i}\right)^{m_i} - 1\right] + \sum_{j=1}^{N_2}\left(\sum_{k=1}^{k}\frac{C_k}{MPC_k} - 1\right) \tag{1.4}$$

where i—for independent individual impurities ($i = \overline{1, N_1}$); j—for synergistic groups of impurities ($i = \overline{1, N_2}$); k—number of impurities in j-th group; m_i—an exponent taking into account the hazard class of the impurity, respectively 1.7—for the first class, 1.3—for the second class, 1—for the third class, 0.8—for the fourth hazard class.

Physical (noise) pollution. An important parameter for assessing the quality of SLAA of UA is physical or noise pollution. The normative and methodological basis for the regulation of adverse effects of noise is a set of state and industry standards and rules, other regulatory and technical documentation on environmental protection from noise exposure, interconnected with a set of standards and rules in the field of environmental protection (or separately atmospheric air) and areas of management of a particular industry (for example, transport) of the national economy.

Environmental and hygienic regulation is carried out in such a way as to satisfy the above requirements and principles of environmental legislation and sanitary-hygienic legislation:

- guarantee an environmentally and hygienically safe acoustic environment for the life and health of the population;
- mandatory compliance with environmental and sanitary-hygienic noise standards (that is, regulatory levels are set in such a way that there is the possibility of their technical, organizational and economic implementation);
- sound coordination of the ecological, economic and social interests of society on the issue of environmental noise;
- scientific justification of the impact of noise on the environment;
- collection of fees for acoustic pollution of the environment, compensation for damage caused by excessive noise levels (for example, costs of measures to strengthen the sound insulation of residential buildings);
- environmental and sanitary-hygienic examination of noise sources and their surroundings.

The decibel scale is based on the Weber-Fechner law [49], according to which the sensitivity of the hearing aid is proportional to the logarithm of the ratio of the intensity of sounds I and I_0.

The logarithmic unit of the ratio of intensities or each step of the scale, which corresponds to a change in the intensity of sound by 10 times, is called in acoustics as Bell (B):

$$L = 10\ \lg\frac{I}{I_0} = 10\ \lg\frac{p^2}{p_0^2} = 20\ \lg\frac{p}{p_0} \qquad (1.5)$$

where I—actual sound intensity, [W/m^2], I_0—threshold level of sound intensity, adopted 10^{-12} W/m^2; P—actual sound pressure, [N/m^2]; P_0—sound pressure of the sensitivity threshold, adopted 2×10^{-5} N/m^2.

According to [50] the equivalent level is the level of constant noise, the effect of which corresponds to the effect of actual noise with variable levels for the same time, measured on the scale "A" of the sound level meter, and is calculated according to formula 1.6:

$$L_{eq}(A) = 10\ \lg\left(\frac{1}{100}\sum_{i=1}^{n} 10^{0.1L_i}t_i\right), \qquad (1.6)$$

where L_i—the average level of class i, [dBA]; t_i—the exposure time of class noise and of the total control time.

The calculated values $L_{eq}(A)$ are compared with normalized sound levels (dBA). For discrete and pulsed noise, acceptable levels are reduced by 5 dB.

For determining the estimated sound level for the day-night period, it can be used the formula 1.7:

$$L_{dn} = 10 \lg\left[\frac{d}{24}10^{0.1(L_d + K_d)} + \frac{24 - d}{24}10^{0.1(L_n + K_n)}\right], \quad (1.7)$$

where L_{dn}—combined (daytime) equivalent sound level for daytime and nighttime, [dBA]; d—duration of the daytime, [h]; L_d—equivalent sound level during the daytime, [dBA] (from 7 to 23 h); L_n—equivalent sound level from night time, [dBA] (from 23 to 7 h); K_d—correction for the weekend (if applicable), [dBA]; K_n—correction for night time, [dBA].

1.4 Determination of Environmental Risk and Methods for Assessing the Risk of Atmospheric Air in UAs

Risk is a very broad concept, reflecting, as a rule, negative phenomena in the life of an individual person or society, the manifestations of which are probabilistic (stochastic) in nature. According to the definition of Knight [51], risk is also a measure of uncertainty that can be determined using probability, that is, situations associated with risk that are characterized by known probabilities of events.

In the work of Dobrovolsky [52], environmental risk is understood as the probability of adverse environmental consequences of any changes in natural objects and factors. Risk is considered as the probability of occurrence of emergency events in a certain period of time, expressed by quantitative parameters. The technogenic aspect of environmental risk is often considered—the likelihood of technological accidents that can cause significant harm to the environment or human health.

Another work [53] gives the concept of environmental risk as the likelihood of negative changes in the environment or the consequences of these changes arising from negative anthropogenic impacts on the environment.

In the textbook on ecology [54], environmental risk is defined as "a danger measure, the degree of possibility of violation of environmental sustainability under any natural or anthropogenic actions leading to exceeding the ecological and economic potential of the nature-territorial complex or object". From this definition it follows that environmental risk factors can be both natural and anthropogenic in nature (although in some cases they partially overlap).

Often environmental risk is interpreted as an indicator of a possible change in the environmental situation with difficult predicting consequences in a particular territory at implementing any type of environmental management.

Ecological risk in this case is presented as a function of two factors: the sustainability (vulnerability) of the landscape to external influences and the level of environmental threat of nature use [55].

So, in [56], the author admits that "environmental risk is understood as the forecast and assessment of the economic damage to the environment from the planned activity, which follows from the environmental assessment of adverse (or dangerous) natural processes and phenomena, optimizing the use of natural resources".

The following definition was given in [57]: "environmental risk is the probability of disturbing ecosystem sustainability, including due to the possible loss of genetic diversity and the occurrence of adverse effects on the life of society (in particular, on public health), as a result of changes in the state of the environment under the influence of anthropogenic and natural factors or as a result of their interaction".

From this definition it is clear that in a generalized form, environmental risk is reduced to two types:

- risk of impaired ecosystem sustainability as a result of real and potential environmental pollution;
- risk to public health, which is the likelihood of adverse effects on health [57].

The given approaches of scientists to the determination of environmental risk demonstrate the ambiguity of this concept, which allows to highlight common features and differences in the definition of this concept, namely:

- environmental risk as a measure of environmental hazard;
- environmental risk as the likelihood of adverse environmental situations;
- environmental risk as a forecast and assessment of economic damage to the environment from the proposed activity.

In the work [57], the authors proceeds from the interpretation of environmental risk as a danger measure, and environmental danger is a situation in which negative events can occur that cause deviations of the state of the environment and the state of human health from their average value; deviation of certain parameters, signs, factors characterizing the state of the environment from their established (optimal, permissible) values.

Moiseev [58] singles out the problem of the "critical point"—a set of critical limitations among the main problems that need to be addressed on the way to constructing a universal methodology for a comprehensive risk analysis for the population and the environment. The author calls the main principle of risk assessment "the principle of critical limit." Thus, it can be evaluated the value of risk as the ratio of expected to critical values.

At the present stage, the concept of risk assessment in almost all countries of the world and international organizations is considered as the main mechanism for developing and making managerial decisions at the international, state, regional levels, at the level of an individual production or other potential source of environmental pollution [57].

To assess the risks arising from chronic exposure to chemicals, annual average concentrations and their upper 95% confidence limits should be applied. During determining the risks of acute (extreme, emergency) situations for a period of up to 24 h, the maximum concentrations are used [30, 33]. Ensuring stable social development, in particular, necessitates the development of tools for measuring the magnitude characterizing the degree of environmental safety in order to determine the permissible anthropogenic load, which should not upset the balance of the environment with the reproduction of its main components and should not cause an increase in the incidence of the population.

The state of the environment may indicate existing threats to environmental problems and risks. The Order of the MHU No. 184 dated April 13, 2007 approved the methodological recommendations "Assessing the risk to public health from air pollution" [59]. These recommendations are intended for specialists of institutions and institutions of the State Sanitary and Epidemiological Service, which assess the level of carcinogenic and non-carcinogenic risks to public health from existing air pollution in the territory of the settlement, which is formed due to industrial emissions, population activity and transformation processes.

By carrying out properly planned epidemiological and environmental hygiene studies, it is possible to identify and quantify the risk of developing diseases associated with the harmful effects of environmental factors for relatively large groups of the population. Today, one of the most effective modern approaches to establishing a connection between the state of the environment and the health of a population in a certain region or city, which allows solving such problems in the conditions of limited time and financial possibilities, is a risk assessment methodology.

A risk assessment methodology is the selection of optimal ways to eliminate or reduce risk in a given situation. It consists of three interrelated elements: risk assessment; risk management; risk communication. It is their combination that allows not only to identify existing problems, to develop ways to solve them, but also to create conditions for the practical implementation of these solutions.

At the same time, determining the risk of air pollution allows to predict the likelihood and medical and social significance of possible health disorders under various scenarios of its impact, but also to establish the priority of measures to manage risk factors at individual and population levels. During determining the risk of exposure to atmospheric air on human health, it is theoretically desirable to take into account the whole spectrum of chemical compounds that can act in this place.

The selection criteria for priority substances of anthropogenic origin are their toxic properties, spread to the environment, resistance, the ability to bio-accumulate and migrate by natural chains, the ability to cause negative effects (irreversible, distant) and the population that they can potentially affect.

In order to determine the level of environmental management hazard and the possibility of environmental emergencies in this book, a new approach to assessing the environmental risk of environmental degradation while maintaining existing trends in anthropogenic pressure is presented. To obtain effective estimates, each risk component must be adapted to problems of an appropriate spatial scale.

The general model of environmental risk assessment is given [60]:

$$M = \{q_p | p = \overline{1, n}\}, \quad (1.8)$$

where M—set of risk factors, and q_p—factor of the p-th type of risk.

Each risk factor q_p is characterized by anthropogenic pressure indicator H_p and natural factors.

Anthropogenic pressure as a cause of ecosystem stability disturbance is associated with the j-th number of indicators h_{pj}, determined by [60]:

$$H_p = \{h_{pj} | p \in N;\ j = \overline{1, n_p}\}, \quad N \in [1, n], \quad (1.9)$$

where H_p—current level of anthropogenic pressure on the j-th component of the environment, which causes the p-type of risk of negative consequences for the ecosystem, N is the number of factors of the p-type risk of exposure to anthropogenic pressure on the j-th component of the environment.

Of course, environmental risk as the probability of disturbance of ecosystem stability depends on the existing state of ecosystem components (K_i) and the influence of modern or potential (possible) anthropogenic pressure (H_i) and can be expressed by the function [60]:

$$P = f_i(K_i, H_i). \quad (1.10)$$

In general, the probability of a violation of the environmental sustainability of the i-th environmental component under the influence of negative factors is determined by the next formula:

$$P = 1 - \prod_{i=1}^{k} (1 - P_i), \quad (1.11)$$

where P_i is the probability of disturbance of ecosystem stability, which depends on the existing state of the i-th component of the ecosystem, K_i is the state of the i-th component of the environment, and H_i is the level of anthropogenic pressure on the i-th component of the environment.

Environmental risk assessment in the current state of the i-th environmental component is determined by the following formula [60]:

$$P_i^c = f\left(K_i^c, H_i^c\right), \quad (1.12)$$

where K_i^c is the current state of the i-th environmental component, H_i^c is the current level of anthropogenic pressure from the impact of negative factors on the i-th environmental component.

The achievement of the critical state (K_i^{cr}) of the i-th component of the environment, in which degradation processes develop and ecosystem stability is disturbed, can occur according to several scenarios.

Firstly, when the current state of the ecosystem is near critical, then even a small anthropogenic pressure (H_i) can lead to intensive development of degradation processes, and secondly, when anthropogenic pressure exceeds the permissible volumes.

1.5 Conclusions

1. An analysis of literary sources has shown that ES of cities are considered as one of the forms of human adaptation to the environment. The growth of cities, the development of industry and transport in them is an objective reality of the modern world. Each UA has its own characteristic technical parameters, which must be taken into account to determine the levels of urban distribution of ES taking into account environmental risks. It has been established that the identification of the main environmental factors in UA is carried out using various methods for monitoring and diagnosing environmental quality, modeling the level of pollution for chemical and physical factors.
2. One of the indicators reflecting the functional state of the urban environment is the criterion for the comfort of living conditions. The research results showed that the direction of changes in environmental safety depends on the density of population in UA, provision of green spaces and greening of UA, development of transport infrastructure, provision of water and recreational resources.
3. It is shown that during assessing the chemical pollution of SLAA, one should take into account the hazard class of pollutants, the frequency of exceeding the MPC of the impurity, the nature of the combined action of substances that are jointly present in the air. It was gained distribution the API of cities. An important parameter for assessing the quality of SLAA of UA is physical pollution (noise).
4. The main approaches to the assessment of environmental risk under the conditions of anthropogenic loads are investigated and the corresponding mathematical models of a statistically assessed category are obtained, which are a vector multicomponent value. During conducting the analysis, it was found that in the environmental safety problem of UA, the laws of the fight against physical pollution of atmospheric air—noise, electromagnetic and ionizing—are not taken into account.

References

1. Gollin, D., Jedwab, R., Vollrath, D.: Urbanization with and without industrialization. J. Econ. Growth **21**, 35–70 (2016). https://doi.org/10.1007/s10887-015-9121-4
2. Guy, M., Rauch, F., Redding, S.J.: Urbanization and structural transformation. Q. J. Econ. **127**(2), 535–586 (2012). https://doi.org/10.1093/qje/qjs003
3. Elmqvist, Th., Fragkias, M., Goodness, J., Güneralp, B., Marcotullio, P.J., McDonald, R.I., Parnell, S., Schewenius, M., Sendstad, M., Seto, K.C., Wilkinson, C. (eds.).: Urbanization, Biodiversity and Ecosystem Services: Challenges and Opportunities. Springer (2013). https://doi.org/10.1007/978-94-007-7088-1
4. Popov, O., Iatsyshyn, A., Kovach, V., Artemchuk, V., Taraduda, D., Sobyna, V., Sokolov, D., Dement, M., Yatsyshyn, T.: Conceptual approaches for development of informational and analytical expert system for assessing the NPP impact on the environment. Nucl. Radiat. Saf. **79**(3), 56–65 (2018). https://doi.org/10.32918/nrs.2018.3(79).09
5. Nikitin, E.D., Skvortsova, E.B., Kochergin, A.N., Nikitina, O.G., Ivanov, O.P., Sabodina, E. P., Vorontsova, E.M.: Development of the concept of the ecological functions of the soil cover and other geospheres. Eurasian Soil Sci. **43**, 721–727 (2010). https://doi.org/10.1134/S106422931007001X
6. Romanukha, D.P., Zatserkovnyi, V.I., Savkov, P.A., Pampukha, I.V., Syniavska, I.K.: Valuation of the impact of urbanization on the geoecological problems of the megapolis. In: Conference Proceedings, Geoinformatics: Theoretical and Applied Aspects, May 2020, vol. 2020, pp. 1–5 (2020). https://doi.org/10.3997/2214-4609.2020geo125
7. Fang, C., Wang, J.: A theoretical analysis of interactive coercing effects between urbanization and eco-environment. Chin. Geogra. Sci. **23**, 147–162 (2013). https://doi.org/10.1007/s11769-013-0602-2
8. Iatsyshyn, A., Popov, O., Artemchuk, V., Kovach, V., Kameneva, I.: The peculiarities of the thermal power engineering enterprise's ash dumps influence on the environment. Prob. Emerg. Situations **28**, 57–68 (2018). https://doi.org/10.5281/zenodo.2594489
9. Iatsyshyn, A., Artemchuk, V., Zaporozhets, A., Popov, O., Kovach, V.: Mathematical approaches for determining the level of impact of ash-slag dumps of energy facilities on the environment. In: Babak V., Isaienko V., Zaporozhets A. (eds) Systems, Decision and Control in Energy I. Studies in Systems, Decision and Control, vol. 298. Springer, Cham (2020). https://doi.org/10.1007/978-3-030-48583-2_1
10. Popov O., Taraduda D., Sobyna V., Sokolov D., Dement M., Pomaza-Ponomarenko A.: Emergencies at potentially dangerous objects causing atmosphere pollution: peculiarities of chemically hazardous substances migration. In: Babak V., Isaienko V., Zaporozhets A. (eds) Systems, Decision and Control in Energy I. Studies in Systems, Decision and Control, vol. 298, pp. 151–163 (2020). https://doi.org/10.1007/978-3-030-48583-2_10
11. Bondarenko, O., Movchan, Y., Gulevets, D., Goutsal, O., Drapaliuk, A.: Establishment of the Chornobyl radioecological biosphere reserve as recovery biotechnology implementation of the exclusion zone. Proc. Natl. Aviat. Univ. **4**(69), 89–95 (2016). https://doi.org/10.18372/2306-1472.69.11060
12. Kaza, N.: The changing urban landscape of the continental United States. Landscape and Urban Plann **110**, 74–86 (2013). https://doi.org/10.1016/j.landurbplan.2012.10.015
13. Kovach, V., Lysychenko, G.: Toxic soil contamination and its mitigation in Ukraine. In: Dent D., Dmytruk Y. (eds.) Soil Science Working for a Living. Springer, Cham (2017). https://doi.org/10.1007/978-3-319-45417-7_18
14. Shkitsa, L.E., Yatsyshyn, T.M., Popov, A.A., Artemchuk, V.A.: The development of mathematical tools for ecological safe of atmosphere on the drilling well area. Neftyanoe khozyaystvo—Oil Industry **11**, 136–140 (2013)
15. Erle, C.E.: Anthropogenic transformation of the terrestrial biosphere. Philos. Trans. R. Soc. A: Math. Phys. Eng. Sci. **369**(1938), 1010–1035 (2011). https://doi.org/10.1098/rsta.2010.0331

16. Maantay, J., Maroko, A.: Mapping urban risk: flood hazards, race, & environmental justice in New York. Appl. Geogr. **29**(1), 111–124 (2009). https://doi.org/10.1016/j.apgeog.2008.08.002
17. Liao, Z., Mao, X., Hannam, P.M., Zhao, T.: Adaptation methodology of CBR for environmental emergency preparedness system based on an improved genetic algorithm. Expert Syst. Appl. **39**(8), 7029–7040 (2012). https://doi.org/10.1016/j.eswa.2012.01.044
18. Popov, O., Iatsyshyn, A., Kovach, V., Artemchuk, V., Taraduda, D., Sobyna, V., Sokolov, D., Dement, M., Hurkovskyi, V., Nikolaiev, K., Yatsyshyn, T., Dimitriieva, D.: Physical features of pollutants spread in the air during the emergency at NPPs. Nucl. Radiat. Saf. **84**(4), 88–98 (2019). https://doi.org/10.32918/nrs.2019.4(84).11
19. Vos, P.E.J., Maiheu, B., Vankerkom, J., Janssen, S.: Improving local air quality in cities: to tree or not to tree? Environ. Pollut. **183**, 113–122 (2013). https://doi.org/10.1016/j.envpol.2012.10.021
20. Lange, E.: Our shared landscape: design, planning and management of multifunctional landscapes. J. Environ. Manage. **89**(3), 143–145 (2008). https://doi.org/10.1016/j.jenvman.2007.06.013
21. Krause, C.L.: Our visual landscape: Managing the landscape under special consideration of visual aspects. Landscape and Urban Plann. **54**(1–4), 239–254 (2001). https://doi.org/10.1016/S0169-2046(01)00139-6
22. Osmond, C.B., Bjorkman, O., Anderson, D.J.: Physiological processes in plant ecology: the structure for a synthesis. In: Physiological Processes in Plant Ecology. Ecological Studies (Analysis and Synthesis), vol. 36, pp. 1–11. Springer, Berlin, Heidelberg (1980). https://doi.org/10.1007/978-3-642-67637-6_1
23. Appleyard, S., Wong, S., Willis-Jones, B., Angeloni, J., Watkins, R.: Groundwater acidification caused by urban development in Perth, Western Australia: source, distribution, and implications for management. Aust. J. Soil Res. **42**(6), 579–585 (2004). https://doi.org/10.1071/SR03074
24. Carmon, N., Shamir, U., Meiron-Pistiner, S.: Water-sensitive urban planning: protecting groundwater. J. Environ. Plann. Manage. **40**(4), 413–434 (1997). https://doi.org/10.1080/09640569712010
25. Wang, J., Chen, L., Zha, J., Ye, Z.: Development of global low-carbon cities. In: Huang W., Wang M., Wang J., Gao K., Li S., Liu C. (eds.) China Low-Carbon Healthy City, Technology Assessment and Practice. Environmental Science and Engineering, pp. 59–90. Springer, Berlin, Heidelberg (2016). https://doi.org/10.1007/978-3-662-49071-6_4
26. Xiao, M., Lin, Y., Han, J., Zhang, G.: A review of green roof research and development in China. Renew. Sustain. Energy Rev. **40**, 633–648 (2014). https://doi.org/10.1016/j.rser.2014.07.147
27. Bach, A., Pawlowska, B., Pietrzak, M.: Eco-friendly methods of reducing the consequences of winter maintenance in urban green areas. Folia Horticulturae **21**(2), 99–109 (2013). https://doi.org/10.2478/fhort-2013-0142
28. Wei, X., Lyu, S., Yu, Y., Wang, Z., Liu, H., Pan, D., Chen, J.: Phylloremediation of air pollutants: exploiting the potential of plant leaves and leaf-associated microbes. Front. Plant Sci. **8**, 1318 (2017). https://doi.org/10.3389/fpls.2017.01318
29. Patterson, P.H.: Management strategies to reduce air emissions: emphasis—dust and ammonia. J. Appl. Poult. Res. **14**(3), 638–650 (2005). https://doi.org/10.1093/japr/14.3.638
30. Zaporozhets, A., Babak, V., Isaienko, V., Babikova, K.: Analysis of the air pollution monitoring system in Ukraine. In: Babak V., Isaienko V., Zaporozhets A. (eds.) Systems, Decision and Control in Energy I. Studies in Systems, Decision and Control, vol. 298, pp. 85–110. Springer, Cham (2020). https://doi.org/10.1007/978-3-030-48583-2_6
31. Weyens, N., Thijs, S., Popek, R., Witters, N., Przybysz, A., Espenshade, J., Gawronska, H., Vangronsveld, J., Gawronski, S.W.: The role of plant-microbe interactions and their exploitation for phytoremediation of air pollutants. Int. J. Mol. Sci. **16**(10), 25576–25604 (2015). https://doi.org/10.3390/ijms161025576

32. Franchini, M., Mannucci, P.M.: Mitigation of air pollution by greenness: a narrative review. Eur. J. Intern. Med. **55**, 1–5 (2018). https://doi.org/10.1016/j.ejim.2018.06.021
33. Isaenko, V., Zaporozhets, A., Babikova, K., Gulevets, D., Savchenko, S.: Review of methods and means on monitoring the air pollution. Proc. Natl. Aviat. Univ. **3**(80), 61–70 (2019). https://doi.org/10.18372/2306-1472.80.14275
34. Harrison, R.M., Struges, W.T.: Physico-chemical speciation and transformation reactions of particulate atmospheric nitrogen and sulphur compounds. Atmos. Environ. (1967) **18**(9), 1829–1833 (1984). https://doi.org/10.1016/0004-6981(84)90358-5
35. Fanizza, C., De Berardis, B., Ietto, F., Soggiu, M.E., Schiro, R., Inglessis, M., Ferdinandi, M., Incoronato, F.: Analysis of major pollutants and physico-chemical characteristics of PM2.5 at an urban site in Rome. Sci. Total Environ. **616–617**, 1457–1468 (2018). https://doi.org/10.1016/j.scitotenv.2017.10.168
36. Krupa, S.V.: Sampling and physico-chemical analysis of precipitation: a review. Environ. Pollut. **120**(3), 565–594 (2002). https://doi.org/10.1016/S0269-7491(02)00165-3
37. Xu, X., Gonzalez, J.E., Shen, S., Miao, S., Dou, J.: Impacts of urbanization and air pollution on building energy demands—Beijing case study. Appl. Energy **225**, 98–109 (2018). https://doi.org/10.1016/j.apenergy.2018.04.120
38. Zaporozhets A.: Research of the process of fuel combustion in boilers. In: Control of Fuel Combustion in Boilers. Studies in Systems, Decision and Control, vol. 287, pp. 35–60. Springer, Cham (2020). https://doi.org/10.1007/978-3-030-46299-4_2
39. Babak V.P., Babak S.V., Myslovych M.V., Zaporozhets A.O., Zvaritch V.M.: Technical provision of diagnostic systems. In: Diagnostic Systems for Energy Equipments. Studies in Systems, Decision and Control, vol. 281, pp. 91–133. Springer, Cham (2020). https://doi.org/10.1007/978-3-030-44443-3_4
40. Zaporozhets, A.: Analysis of control system of fuel combustion in boilers with oxygen sensor. Periodica Polytech. Mech. Eng. **63**(4), 241–248 (2019). https://doi.org/10.3311/PPme.12572
41. Cerra, J.F.: Emerging strategies for voluntary urban ecological stewardship on private property. Landscape and Urban Plann. **157**, 586–597 (2017). https://doi.org/10.1016/j.landurbplan.2016.06.016
42. Yokohari, M., Amemiya, M., Amati, M.: The history and future directions of greenways in Japanese New Towns. Landscape and Urban Plann. **76**(1–4), 210–222 (2006). https://doi.org/10.1016/j.landurbplan.2004.09.033
43. Kim, G.: The public value of urban vacant land: social responses and ecological value. Sustainability **8**(5), 486 (2016). https://doi.org/10.3390/su8050486
44. Chen, Z.-L., Chen, J.-Y., Liu, H., Zhang, Z.-F.: Present status and development trends of underground space in Chinese cities: evaluation and analysis. Tunn. Undergr. Space Technol. **71**, 253–270 (2018). https://doi.org/10.1016/j.tust.2017.08.027
45. Li, H.-Q., Parriaux, A., Thalmann, P., Li, X.-Z.: An integrated planning concept for the emerging underground urbanism: deep city method part 1 concept, process and application. Tunn. Undergr. Space Technol. **38**, 559–568 (2013). https://doi.org/10.1016/j.tust.2013.04.010
46. van der Straaten, J.: Overview of the case studies. In: Nowicki, P., Bennet, G., Middleton, D., Rientjes, S., Wolters, R. (eds.) Perspectives on Ecological Networks, pp. 121–131. European Centre for Nature Conservation (1996)
47. Vuilleumier, S., Prelaz-Droux, R.: Map of ecological networks for landscape planning. Landscape and Urban Plann. **58**(2–4), 157–170 (2002). https://doi.org/10.1016/S0169-2046(01)00218-3
48. Wackernagel, M., Rees, M.: Let's just assume we're sustainable. Ecol. Econ. **211**, 3–24 (1997)
49. Portugal, R.D., Svaiter, B.F.: Weber-Fechner law and the optimality of the logarithmic scale. Mind. Mach. **21**, 73–81 (2011). https://doi.org/10.1007/s11023-010-9221-z
50. Pandya, G.H.: Assessment of traffic noise and its impact on the community. Int. J. Environ. Stud. **60**(6), 595–602 (2003). https://doi.org/10.1080/0020723032000093973

51. Knight, F.H.: Selected Essays by Frank H. Knight, Volume 2: Laissez Faire: Pro and Con (vol. 2). University of Chicago Press (1999)
52. Dobrovolsky, V.V.: Fundamentals of the theory of ecological systems. Kyiv, Professional (2005). (in Ukrainian)
53. Jones, R.N.: Explanatory dictionary, an environmental risk assessment/management framework for climate change impact assessments. Nat. Hazards **23**, 197–230 (2001). https://doi.org/10.1023/A:1011148019213
54. Solukha, B.V., Fuks, G.B.: Urban Ecology: A Textbook. KNUBA, Kyiv (2004)
55. Maurin, A.M.: The principle of irreversibility in modeling bioecological systems. In: Modeling and Projection in Ecology. University of Latvia, Riga (1978). (in Russian)
56. Lysychenko, G.V., Zabulonov, YuL, Khmil, G.A.: Natural, man-made and environmental risks: analysis, assessment, management. Naukova dumka, Kyiv (2008). (in Ukrainian)
57. Lysychenko, G.V., Khmil, G.A., Barbashev, S.V.: Methodology for environmental risk assessment. Astroprint, Odesa (2011). (in Ukrainian)
58. Moiseev, N.N.: Mathematical problems of system analysis, vol. 488. Nauka, Moscow (1981). (in Russian)
59. On risk assessment for public health from air pollution: Order of the Ministry of Health of Ukraine No. 184 of 13.04.2007/Ministry of Health and Health of Ukraine, Official Gazette of Ukraine, Kyiv No. 1 dated 21 April 2007
60. Rybalova, O.V., Belan, S.V., Artemiev, S.R.: The definition of environmental risk deterioration of air considering the chemical danger of the regions of Ukraine. Probl. of emergencies **18**, 196–203 (2013). (in Ukrainian)

Chapter 2
Improvement the Methods for Assessing the Hazard of the Surface Layer of Atmospheric Air and the Ecological Risk of the State of Urban Areas

Abstract In this section, a study of the complex influence of physical and chemical factors on the quality of the surface layer of atmospheric air is carried out, as the basis for the existence and ensuring the ecological safety of an urbanized territory. Since the indicators of physical and chemical pollution are expressed in absolute and relative values, their values must be reduced to the same dimension. For this, it is proposed to apply the method of assessing the air quality index (AQI). In contrast to the above method, where the corresponding concentrations of the pollutant are taken outside the interval, it is proposed to use the intervals of the multiplicity of the maximum permissible concentration's excess of pollution to their standard values, including the determination of the level of pollution and the degree of its hazard, respectively.

Keywords Chemical and physical factors · Pollution · Air quality index · Hazard levels · Complex impact · Urban areas

2.1 Algorithm and Stages of Studying the State of the Environment of UAs

Environmental studies were carried out on the territory of the capital of Ukraine—Kyiv. The general scheme of research organization is shown in Fig. 2.1.

The study included such physical and chemical factors of impact on UA, obtained in the course of own environmental studies (noise pollution, electromagnetic pollution and pollution from consumed energy), as well as studies of the CGO in Kyiv and scientists [1–3], presented in Table 2.1.

The methodological stages of the study are shown in Fig. 2.2.

D. Gulevets et al., *Environmental Control for Ensuring Cities Safety*, Lecture Notes in Intelligent Transportation and Infrastructure, https://doi.org/10.1007/978-3-030-66710-8_2

Purpose		assessment of the role of technogenic factors in the formation of the environmental hazard of population living in the city
	⇩	
Object		influence of physical and chemical factors on atmospheric air
	⇩	
Information source		own noise measurements; statistical information of the CGO of the Kyiv city, reports on the state of the environment of the Kyiv city, scientific results of previous researchers
	⇩	
Methods	**Ecological and analytical**	noise observations
	Geoinformation and analytical	creation of electronic databases; geoinformation mapping
	Mathematical and statistical	mathematical and statistical analysis and modeling (calculation of hazard indices for atmospheric air, the risk of exposure to polluted atmosphere on the population of UA)

Fig. 2.1 General scheme of the research organization

2.2 Method for Assessing the Hazard of the Surface Layer of Atmospheric Air in an UA

This method proposes to evaluate one of the factors of the environmental criterion for assessing the environmental safety of UA. The studies were aimed at developing an assessment of the environmental hazard of SLAA of UA, which is created by modern anthropogenic loads under the influence of chemical and physical pollution factors.

It is proposed to determine the hazard of atmospheric air according to the following algorithm (Fig. 2.3).

2.2.1 Determination of Priority Factors of Pollution of the Surface Layer of Atmospheric Air in UA

SLAA pollution is the introduction of new uncharacteristic physical, chemical and biological substances into the atmospheric air or changes in their natural concentration.

Table 2.1 Chemical and physical research data

Research factors	Own research	Previous research data
Chemical pollution		
Atmospheric air	–	CGO data: the concentration of dust, sulfur dioxide, sulfates, carbon monoxide, nitrogen dioxide, nitrogen oxide, hydrogen sulfide, ammonia, phenol, formaldehyde, hydrogen chloride and fluoride; heavy metals concentrations: iron, cadmium, lead, zinc, copper, nickel, manganese, chromium
Condition of green spaces	–	Data of M.G. Kholodny Institute of Botany
Physical pollution		
Noise pollution	Assessment of the state of noise pollution from air transport	–
Electromagnetic pollution	–	Results of determining the levels of electromagnetic pollution and its load on the population of the administrative districts of the Kyiv city
Consumed energy	–	Data of the statistical collection of Ukraine for 2013: coal, natural gas, motor gasoline, heavy fuel oil, firewood for heating, fuel briquettes

According to pollution sources, two types of atmospheric air pollution are distinguished: natural, artificial. By the nature of the pollutant, atmospheric pollution is of three types:

- Physical—mechanical (dust, solid particles), radioactive (radiation and isotopes), electromagnetic (different types of electromagnetic waves, including radio waves), noise (various loud sounds and low-frequency vibrations) and thermal pollution (for example, emissions of warm air, etc.);
- Chemical—pollution by gaseous substances and aerosols. The main chemical air pollutants are carbon monoxide (IV), nitrogen oxides, sulfur dioxide, hydrocarbons, aldehydes, heavy metals (Pb, Cu, Zn, Cd, Cr), ammonia, atmospheric dust and radioactive isotopes [1];
- Biological—mainly microbial pollution. For example, air pollution by vegetative forms and spores of bacteria and fungi, viruses, their toxins and waste products [4, 5].

The main pollutants are carbon monoxide; nitrogen oxides sulfur dioxide; hydrocarbons; aldehydes; heavy metals (Pb, Cu, Zn, Cd, Cr) ammonia; atmospheric dust; radioactive isotopes [1].

Carbon monoxide (CO) is a colorless, odorless gas. It is formed as a result of incomplete combustion of fossil fuels (coal, gas, oil) in conditions of lack of oxygen

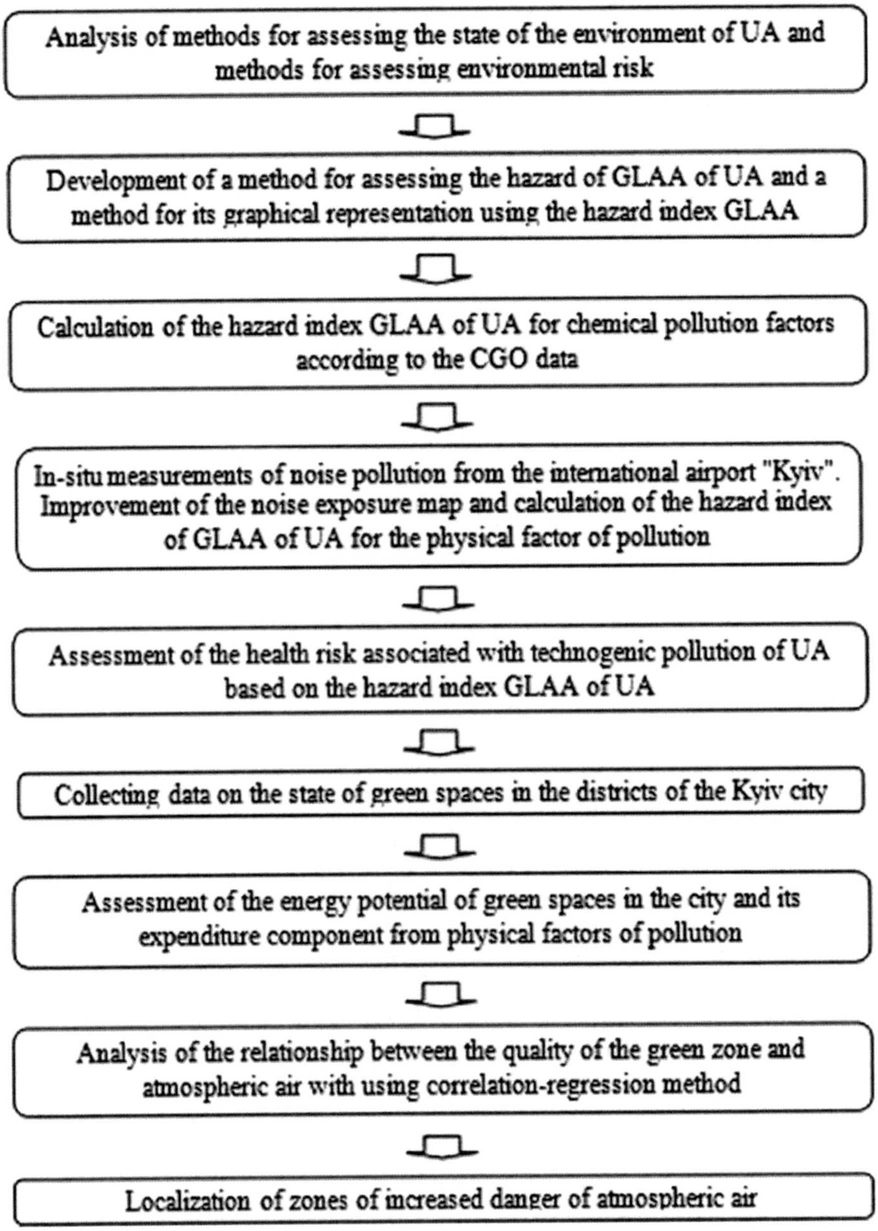

Fig. 2.2 Main stages of the research

and at low temperatures. During inhalation of carbon monoxide, due to the double bond in its molecule, forms strong complex compounds with the hemoglobin of human blood and thereby blocks the flow of oxygen into the blood.

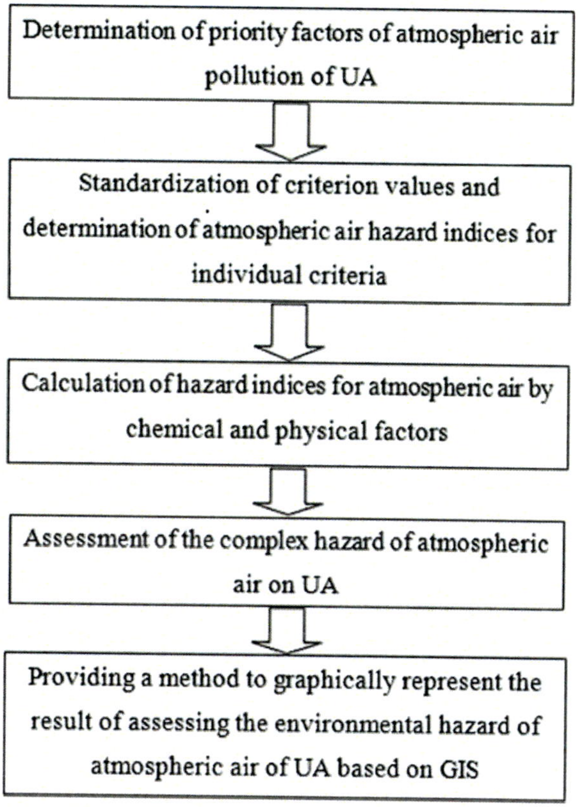

Fig. 2.3 Algorithm for determining the hazard of atmospheric air, taking into account chemical and physical factors of pollution

Carbon dioxide (CO_2) or carbon dioxide is a colorless gas with a sour smell and taste, a product of the complete oxidation of carbon. It is one of the greenhouse gases.

Sulfur dioxide (SO_2) is a colorless gas with a pungent odor. It is formed during the combustion of sulfur-containing fossil fuels, mainly coal. Global SO_2 emissions are estimated at 190 million tons per year. Long-term exposure to sulfur dioxide on humans and warm-blooded animals first leads to loss of taste, limited breathing, inflammation or edema of the lungs, interruptions in cardiovascular activity, circulatory disorders and respiratory arrest [1, 6, 7].

Nitrogen oxides (nitrogen monoxide and nitrogen dioxide) are gaseous substances: nitrogen monoxide NO and nitrogen dioxide NO_2 are combined by one common formula NO_x. The higher the combustion temperature, the more intensive is the formation of nitrogen oxides [1, 8–10].

Another source of nitrogen oxides are enterprises producing nitrogen fertilizers, nitric acid and nitrates, aniline dyes, and nitro compounds. The amount of nitrogen oxides released into the atmosphere is 65 million tons per year. Transport accounts for 55% of the total amount of nitrogen oxides emitted into the atmosphere, energy —28%, industrial enterprises—14%, small consumers and the household sector—3% [1, 11].

Ozone (O_3) is a gas with a characteristic odor, a stronger oxidizing agent than oxygen. It is ranked among the most toxic of all common air pollutants. Ozone is formed in the lower atmospheric layer as a result of photochemical processes involving nitrogen dioxide and light organic compounds [12–14].

Hydrocarbons are chemical compounds of carbon and hydrogen. These include thousands of various substances that pollute the atmosphere, which are contained in products of incomplete combustion of gasoline, liquids used in dry cleaning, industrial solvents, etc. [15–17].

Plumbum (Pb) is a silvery-gray metal that is toxic in any known form. Widely used for the production of paints, ammunition, printing alloy and the like. About 60% of the world's plumbum production is used annually for the production of acid batteries.

However, the main source (about 80%) of air pollution from plumbum compounds is in the exhaust gases of vehicles that use leaded gasoline [18].

Industrial dust, depending on the mechanism of formation, is divided into the following 4 classes:

- mechanical dust—formed as a result of grinding the product during the technological process;
- distillation—formed as a result of volumetric condensation of a vapor of substances during cooling of gas passed through a technological apparatus, installation or unit;
- light ash—contained in the flue gas in suspension in the fuel residues, does not burn, it is formed from its mineral impurities during combustion;
- industrial soot—part of industrial emissions, solid highly dispersed carbon, formed during incomplete combustion or thermal decomposition of hydrocarbons.

2.2.2 *Standardization of Criterion Values and Determination of Atmospheric Air Condition Indices According to Individual Criteria*

Since the indicators of physical and chemical pollution are expressed in absolute and relative values and have different units of measurement, then at the second step it is necessary to bring the values to the same dimension. For this it is proposed to use the AQI method, which is used by the United States Environmental Protection Agency to report air quality [1, 3]. This index is a tool for providing information on ambient air pollution to the general public. The air quality index is calculated based on the concentration indices of five pollutants: O_3, suspended particles, CO, SO_2 and NO_2. AQI is divided into six categories, indicating the level of hazard of atmospheric air to public health [1, 3].

In contrast to the AQI method, where the corresponding levels of the pollutant are taken beyond the boundaries of the interval, it is proposed to use the intervals of the multiplicity of the excess of the air pollution indicators of their standard value, including the determination of the level of pollution and the degree of its danger (Table 2.2).

The atmospheric air hazard index is proposed to be calculated by the formula:

$$I_{haa} = \frac{(I_u - I_l)}{(M_u - M_l)} \times \left(K_p - M_l\right) + I_n, \tag{2.1}$$

where I_{haa}—value of the SLAA hazard index; K_p—multiplicity of exceeding the normative value of the MPC; M_u—limit of exceeding the multiplicity of the normative MPC value, greater than or equal to K_p; M_l—limit of exceeding the multiplicity of the normative MPC value, less than K_p; I_u—value of the quality index corresponding to M_u; I_l—value of the quality index corresponding to M_l.

In contrast to the AQI method, where the corresponding concentrations of the pollutant are taken outside the interval, it is proposed to use the intervals of the multiplicity of the excess of the MPC indicators of pollution to their standard value, including determining the level of pollution and the degree of its hazard, respectively.

For determining the level of noise pollution, it is proposed to use the level of noise intensity in terms of the equivalent noise level (LA_{eq}). For the multiplicity of the excess of K_p, the equivalent level obtained in field measurements should be taken.

If an excess of concentration for more than one pollutant is recorded at a meteorological observation post, then the quality of the SLAA is determined by the maximum hazard index of the PLAA.

Table 2.3 shows the boundaries and multiplicity of the pollution level for chemical and physical factors. In accordance with the value of the SLAA hazard index, an appropriate description is assigned.

Explaining for Table 2.3:

- "Safe level" (0–85 points)—fresh air, ideal for relaxation.
- "Moderate level" (>85–170 points)—SLAA quality is acceptable. Some pollutants can be a moderate health problem for a small portion of the population.

Table 2.2 Scale for assessing the level of hazard of atmospheric air

Pollution level	Degree of danger	Excess multiplicity
Acceptable	Healthy	<1
Invalid	Weakly dangerous	>1–2
Invalid	Moderately dangerous	>2–4.4
Invalid	Dangerous	>4.4–8
Invalid	Very dangerous	>8

Table 2.3 Scale of boundaries and multiplicity of pollution levels for chemical factors

Score	State	Multiplicity of exceeding the standard value	Pollution level
0–85	Safe level	<1	Acceptable
>85–170	Moderate level	1–2	Invalid
>170–255	Dangerous level for a sensitive population	>2–4.4	Invalid
>255–425	Dangerous level for all population groups	>4.4–8	Invalid
>425	Very dangerous level	>8	Invalid

- "Dangerous level for a sensitive population" (>170–255 points)—the level of pollution at which the majority of the population will not be affected, but people with lung disease, the elderly and children will be exposed to greater harmful effects.
- "Dangerous level for all population groups" (>255–425 points)—the level of pollution at which everyone can begin to experience some adverse health effects. People in sensitive populations can have serious consequences.
- "Very dangerous level" (>425 points)—this level of pollution can have serious health consequences.

If an excess for more than one pollutant is recorded at the PMP, then the ES of SLAA is determined by the maximum index.

To simplify the calculations, it needs to select the maximum value of the atmospheric air hazard index at the measurement point, the regulatory boundaries of which exceed the MPC.

Noise exceeding sanitary standards has a negative effect on the human body. The effect on the human body of noises that arise in everyday life depends largely on their intensity and nature. Sustained constant noise affects the human body less than noise that occurs irregularly, and noise of a lower frequency is less than high-frequency noise. The latter also contributes to the rapid onset of a person's feeling of fatigue, since low-frequency intense noises have a less pronounced unpleasant effect than high-frequency noise of lower intensity [19–21].

Observations have shown [22] that noise with an intensity level over 60 dBA can inhibit normal stomach digestion, and at a noise of 80–90 dBA the number of stomach contractions per minute decreases by 37%. It was found that at a noise intensity of more than 60 dBA, the secretion of saliva and gastric juice decreases by 44%. A temporary and sometimes permanent increase in blood pressure, increased irritability, decreased performance, mental depression are the consequences of noise. Vague noises that do not reach consciousness also cause depletion of the central nervous system, as a result of which they can cause imperceptible disturbances in the body [23].

The effect of noise on hearing aids is the subject of much research. If a person's hearing aid is exposed to prolonged or repeated exposure to sufficiently loud noise, temporary or permanent hearing loss occurs [22]. A person who is exposed to noise with an intensity of 90 dBA for 6–8 h has a noticeable hearing loss, which disappears approximately 1 h after it stops. Noise exceeding 120 dBA very quickly causes fatigue in a person, which occurs within a few minutes and is accompanied by a noticeable hearing loss. In each case, the degree of hearing loss and the length of the recovery period are proportional to the level of intensity and duration of the effect. At a noise level of 140 dBA (pain threshold) and above, the sensation of pressure increases and spreads throughout the body, and the chest, muscles of the legs and arms begin to vibrate [22].

When the noise intensity reaches 160 dBA, a rupturing of eardrum may occur. To determine the degree of noise pollution, it is proposed to use the noise intensity level in terms of the equivalent noise level (LA_{eq}). It is recommended to take the equivalent level obtained in field measurements as the multiplicity of the excess K_p.

Table 2.4 shows the scale of boundaries and multiplicity of the noise level of pollution. According to the value of the SLAA hazard index score, an appropriate description is assigned.

Explaining for Table 2.4:

- "Good" (0–85 points). Does not cause harmful effects on a person in the process of his life.
- "Moderate" (>85–170 points). It can cause psychological effects, manifested in the deterioration of mental activity, weakening of attention, reaction speed, difficulty in working with arrays of information, and the like.
- "Unsatisfactory for a sensitive population" (>170–255 points). Perhaps its physiological effect: the pulse accelerates, the blood pressure rises, the vessels narrow, which impairs the supply of blood to the organs.
- "Unsatisfactory" (>255–425 points). It can lead to functional disorders in the organs and systems of the human body: auditory sensitivity decreases, the activity of the stomach and intestines worsens, a feeling of nausea, headache, tinnitus appears.
- "Very unsatisfactory" (>425 points). A mechanical effect on the organ of hearing is carried out, which is expressed in a violation of the connections between individual parts of the inner ear, even a rupture of the tympanic

Table 2.4 Scale of boundaries and multiplicity of noise pollution level

Score	State	Sound pollution levels, LA_{eq} (dBA)	Long-term effect
0–85	Good	0–50	None
>85–170	Moderate	>50–70	Intense and irritating effect
>170–255	Unsatisfactory for a sensitive population	>70–90	Hearing hazard possible
>255–425	Unsatisfactory	>90–120	Hearing hazard (duration—8 h)
>425	Very unsatisfactory	>120	Sensitivity threshold

Table 2.5 Description of the state and levels of pollution according to the SLAA hazard index

State	Pollution level	Description
Safe level	Acceptable	SLAA quality is considered satisfactory
Moderate level	Invalid	The quality of SLAA is acceptable. Certain pollutants can be a mild health problem for a very small number of people
Dangerous level for a sensitive population	Invalid	While the majority of the population is likely to be unaffected, people with lung disease, the elderly, and children will be at greater risk of exposure while being in the area
Dangerous level for all population groups	Invalid	Everyone can begin to feel some adverse health effects. People in a sensitive population can have serious consequences
Very dangerous level	Invalid	At this level of pollution, everyone can have serious health consequences

membrane is possible. Such high noise levels affect not only the hearing organs and the entire body. Sound waves, penetrating the skin, cause mechanical vibrations of the body tissues, resulting in the destruction of nerve cells, rupture of small vessels, and the like.

In accordance with the value of the SLAA hazard index, a corresponding description is assigned (Table 2.5).

2.2.3 *Method of Displaying the Hazard Index of the Surface Layer of Atmospheric Air*

The methodology for creating information resources, designing and filling the information DB is based on the systematic principle of organizing the information environment, using unified territorial-sectoral databases and a spatially-oriented system for collecting, accumulating, processing and presenting data. It defines the system-wide principles of base formation, regulates the requirements for the structure, composition and differentiation of data, determines the models of their organization, presentation forms, main sources of information and ways to improve them.

A number of technical solutions are known that make it possible to assess the level of SLAA pollution. They are based on the use of various parameters by which the pollution level [1, 3, 24–27] is estimated.

For example, there is a known method for assessing the degree of anthropogenic air pollution using phytoindication, in particular, by the percentage of affected plant tissue [28].

The main significant disadvantage of this method is its low efficiency. Morphological and structural rearrangements appear in plants under the influence of pollution not immediately, but after a long period of time.

Also a method is known for identifying zones of chemical pollution of SLAA [29], according to which the research area is divided into a conditional grid of squares. The size of the squares is determined depending on how detailed information about the state of the environment is necessary to obtain. Using a biosensor based on a suspension of motile microorganisms, the toxicity index of the sample is determined when its value is less than the minimum permissible or greater than the maximum permissible value of the natural background of the area.

The disadvantage of this method is the large variability of the values obtained, associated with the use of a biosensor based on a suspension of motile microorganisms, large time and financial costs of research.

There is also a method to determine SLAA pollution by chemical analysis of air samples taken at individual points, with subsequent interpolation over the entire area of the controlled territory [30, 31]. In this method, the content of the individual ingredients is first determined and compared with the corresponding MPCs. The use of this method is envisaged in the analysis of field measurements obtained during the monitoring of SLAA of UA, which, due to the limited number of observation posts, does not provide a holistic spatial representation of SLAA pollution, which does not always allow to correctly assess the exposure of pollution in areas remote from the PMP.

The disadvantage of this method is the large material costs for maintaining the network of observation points, the limited information received by the administrative boundaries of the controlled territory, low monitoring efficiency, since a lot of time passes from the moment of sampling to the presentation of the analysis results.

A method is proposed for representing the influence of SLAA pollution based on the SLAA hazard index. This method allows to build a map-schematic of the spatial distribution of the SLAA pollution indices, by which it is possible to quantify the level of pollution of the SLAA of UA. This method of displaying the SLAA hazard index is carried out in the following steps:

1. The UA is selected for which it is necessary to build a SLAA hazard map.
2. On the specified territory, according to n stationary observation posts are located, depending on the population of the city (Table 2.6), on which in situ measurements of the concentrations of pollutants in the SLAA are carried out with the fixation of the date, time of air sampling and meteorological characteristics of the SLAA at the moment of taking the specified sample.
3. The map of the researched UA is covered with a regular grid (Fig. 2.4) with a step of $m \times m$ [m] (the step of the grid is determined by the required final accuracy of the analysis results).
4. The location of n reference points is marked on the map, that is, those at which instrumental full-scale measurements of the quality of the SLAA were performed according to the priority of chemically hazardous substances and physical pollution.

Table 2.6 Number of stationary posts depending on the population (according to Ukraine's law)

Number of posts	Population
1	To 50,000
2	From 50,000 to 100,000
2–3	From 100,000 to 200,000
3–5	From 200,000 to 500,000
5–10	From 500,000 to 1,000,000
10–20 (stationary and route)	From 1,000,000

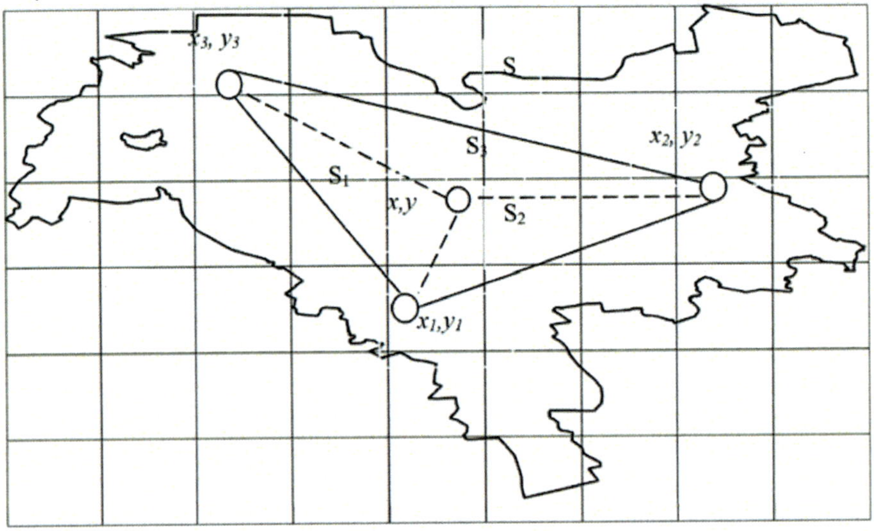

Fig. 2.4 Map with regular grid and triangle system

5. Next, data is collected:
   - coordinates x and y of the location of the PMPs for pollution;
   - characteristics and parameters of SLAA pollution by chemical and physical parameters.
6. The SLAA hazard index is calculated.
7. At n reference points (points of full-scale measurements), the maximum value of the SLAA hazard index at the measurement point is determined.
8. For interpolating the coefficient of conformity of atmospheric air pollution based on the data at the PMP points, a triangulation procedure is carried out. For this, the points of full-scale measurements are combined on a map with a regular grid into triangles, on the basis of which the Delaunay triangulation method [32, 33] is used and the correspondence coefficients are determined at all nodes of the regular computational grid. In more detail, this definition is

carried out as follows. Using the Delaunay triangulation method, all points of full-scale monitoring (points of full-scale measurements) are connected with segments in triangles so that a new segment cannot be added without intersection with the existing segments, thus obtaining a system of triangles on the map.

9. For each triangle, the equations of the plane are solved with the setting of the coefficients of the equation, depending on the coordinates x and y of the vertices of the triangle (points of full-scale measurements) and the values of the coefficients of conformity of the SLAA pollution in them. The plane equation is a continuous linear function of two variables, which can be written in the following form:

$$K(x, y) = a_0 + a_1 x + a_2 y, \tag{2.2}$$

where a_0, a_1, a_2 are arbitrary constant coefficients; $K(x, y)$ is a coefficient of conformity of SLAA pollution for a point with coordinates x, y.
The value of the function at the vertices of the triangle corresponds to the values of the correspondence coefficient at the posts that make up this triangle, and are denoted as k_1, k_2, k_3.
Using the coordinates of the triangle vertices and the values k_1, k_2, k_3, unknown constant coefficients a_0, a_1, a_2 are calculated.
As a result of the above mathematical operations, a system of three linear algebraic equations is formed for the unknown coefficients a_0, a_1, a_2:

$$k_i = K(x_i, y_i) = a_0 + a_1 x_i + a_2 y_i, \quad i = \overline{1, 3}, \tag{2.3}$$

where i—number of the triangle's top, $K(x_i, y_i)$—coefficient of conformity of the SLAA pollution for a point with coordinates x_i, y_i.
By solving Eq. (2.3), it can be obtained an unambiguous expression of function (2.2) through its nodal values (values at grid points) for each triangle.
Next, the attachment of each nodal point located inside the specified system of triangles is determined to one of a series of n obtained triangles according to the following algorithm:

(a) the specified point, located inside the specified system of triangles, is connected by segments with the tops of the each triangles to form three triangles with area S_1, S_2, S_3 (Fig. 2.4).
Moreover, if the area of the original triangle S is equal to the sum of the areas of the formed three triangles $S = S_1 + S_2 + S_3$, where S_1—area of triangle 1; S_2—area of triangle 2; S_3—area of triangle 3, then it is considered that the point belongs to this triangle.

(b) If $S < (S_1 + S_2 + S_3)$, then this point does not belong to this triangle.

The area of triangles is calculated using the following formulas:

$$S = \frac{|(x_2 - x_1) \times (y_3 - y_1) - (x_3 - x_1) \times (y_2 - y_1)|}{2}, \quad (2.4)$$

$$S_1 = \frac{|(x - x_1) \times (y_3 - y_1) - (x_3 - x_1) \times (y - y_1)|}{2}, \quad (2.5)$$

$$S_2 = \frac{|(x_2 - x) \times (y_3 - y) - (x_3 - x) \times (y - y_2)|}{2}, \quad (2.6)$$

$$S_3 = \frac{|(x_2 - x_1) \times (y - y_1) - (x - x_1) \times (y_2 - y_1)|}{2}, \quad (2.7)$$

where x, x_1, x_2, x_3—coordinates of the triangle's top along the X axis; y, y_1, y_2, y_3—coordinates of the triangle's top along the Y axis.
Next, the value of the match coefficient is calculated at the nodes of the regular grid, inside each triangle using the obtained coefficients for the corresponding triangle.

10. Extrapolation of the values of the correspondence coefficient of the nodal points lying outside the obtained system of triangles is carried out. The value of the coefficient for these points is taken equal to the coefficients at the nearest point lying on the border of the system of triangles formed by the points of full-scale measurements (rapper points).
11. As a result of these actions, the approximated value of the coefficient of pollution's conformity with the SLAA at all nodes of the regular grid is obtained.
12. Next, a map-diagram of the spatial distribution of atmospheric air pollution indices is constructed, according to which the level of atmospheric air pollution in the study area is quantified.

2.3 Method for Assessing the Environmental Risk of the Surface Layer of Atmospheric Air Under Conditions of Technogenic Loads

At the present stage, the concept of risk assessment in almost all countries of the world and international organizations is considered as the main mechanism for the development and adoption of managerial decisions at the international, state, regional levels [34].

It should be noted that the study of risks is most often directly related to the study of the expected impact of the most dangerous natural and man-made processes.

In recent years, more and more attention of researchers has been attracted by the issues of studying environmental risk in connection with its insurance [34, 35]. Thus, during studying environmental risk, the assessment of the existing state of natural-territorial complexes, as a rule, fades into the background. Meanwhile, the determination of the degree of technogenic changes in various natural components under the conditions of the existing technogenic impact is a very important and urgent task.

Environmental risk is considered from a dual standpoint, therefore, the complex state of the environment is assessed based on the results of chemical and physical factors affecting the environment. It is proposed to use the SLAA system of hazard indexes as environmental criteria. The following methodology is proposed for calculating the determination of individual risk, which depends on the level of danger of SLAA for the population and the probability of exceeding the standards for atmospheric air quality.

The method can be used to calculate the individual risk during coordinating the construction and design of city development in order to regulate the life of people (population and workers), which are daily exposed to the risk of deteriorating health or getting diseases in case of pollution, which exceeds the SLAA quality standards.

Calculation of the individual risk of SLAA danger consists of the following steps:

1. Determination of the annual number of observations for each species. The total number of observations *N*, which are investigated, is calculated as the sum of observations of each type of pollution for the observation point:

$$N = N_1 + N_2 + \cdots + N_n = \sum_{i=1}^{n} N_n, \qquad (2.8)$$

   where *i*—SLAA pollution category, *n*—amount of the type of pollution, *N*—number of observations.
2. Average weighted estimate of the amount of exceeding the permissible hazard limit for all types of SLAA pollution is calculated. This calculation is carried out using an estimate of the frequency of exceeding the permissible limit—the number of times the pollution rate is exceeded per the total number of observations of each type of pollution for each type of SLAA pollution.
3. The probability of exceeding the permissible hazard limit for certain types of SLAA pollution is calculated. The average value of the total number of observations during the year and the value of the assessment of exceeding the permissible hazard limit of atmospheric air pollution R (for typical chemical or physical SLAA pollution), which is calculated as a weighted average for all types of pollution, are determined for each PMP:

$$P_m = \frac{(R_1 N_1 + R_2 N_2 + \cdots + R_n N_n)}{N} = \frac{\sum_{i=1}^{n} R_n N_n}{N}, \qquad (2.9)$$

where P_m—probability of exceeding the permissible hazard limit for certain types of SLAA pollution, R—frequency of exceeding the permissible hazard limit for certain types of pollution, N—number of observations.

4. The probability of atmospheric air pollution by levels of excess of the pollution state is calculated.
It is known that one of the most effective risk assessment models is the probabilistic model. In [34, 35], it was proposed as a criterion that provides an average estimate of the probability of the degree of environmental pollution, to use the integral of the distribution function of pollutant concentrations, taken within the MPC of each substance to the maximum concentration value. Despite this and taking into account that the law of distribution of harmful elements in the SLAA is often close to normal (symmetric), for approximating the probability density of the distribution of harmful contents it was proposed to use the Weibull distribution, which allows to describe the main types of distributions (from right-asymmetric and left-asymmetric).
Based on this provision, the criterion for assessing the probability of the risk of contamination by the j-th substance is:

$$P = \exp\left[-\left(\frac{b_j \times I}{S_j}\right)^{m_j}\right], \quad (2.10)$$

where b_j and m_j—the Weibull distribution parameters for the j-th harmful substance; S_j—average value of the concentration of the j-th harmful substance in the environment of a given (estimated) time interval.

5. According to [1, 3, 36], the ecological state is classified according to the increasing degree of danger: safe level, moderate level, dangerous level for a sensitive group of the population, dangerous level and very dangerous level. Suppose that the distribution density function of the pollution index of the components of the environment is known (Fig. 2.5), which it was expressed through $\varphi(Z)$. The ecological state of the environment, depending on the value of Z, will correspond to:

- safe level if $Z < Z_1$;
- moderate level if $Z_1 < Z < Z_2$;
- dangerous level for a sensitive group of the population, if $Z_2 < Z < Z_3$;
- dangerous level $Z_3 < Z < Z_4$;
- very dangerous level $Z > Z_4$.

The integral from the distribution density function in the range from 0 to ∞ is the area of the distribution density curve Z, which equal to 1. Obviously, the integral taken in the range from Z_1 to $+\infty$, that is, in the range of values that exceeds the safe level equal to Z_1, will be represent the probability of a mediocre risk of contamination.

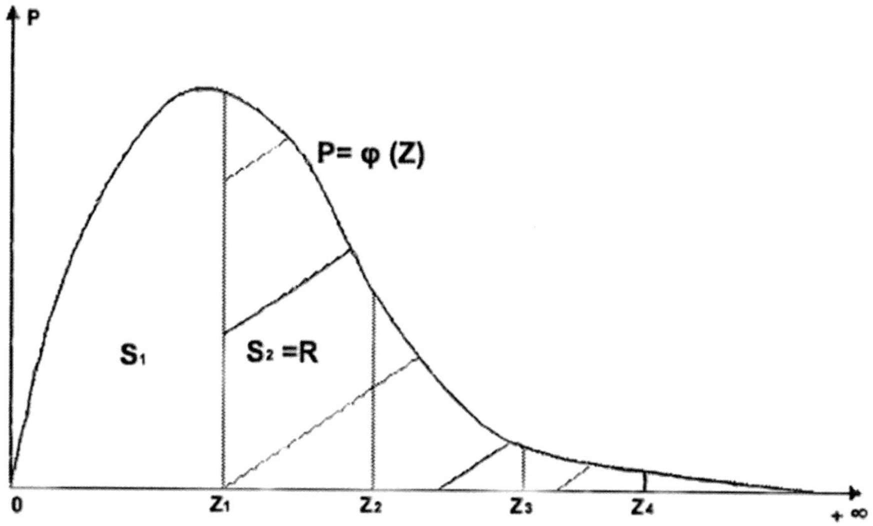

Fig. 2.5 Graph of the distribution density function of the environmental pollution indicator

Graphically this is the shaded area S_2 under the curve of the distribution density function $P = \varphi(Z)$, where $S_2 < 1$, and $S_1 + S_2 = 1$.
Accordingly, the limits of integration from Z_2 to $+\infty$ determine the risk of exceeding the dangerous level for a sensitive group of the population. The range of integration from Z_3 to $+\infty$ determines the probability of the dangerous pollution level, and the integral in the range from Z_4 to $+\infty$ determines the risk of a very dangerous level of pollution of the SLAA.
In general, the ecological probability of the risk of hazardous pollution of the SLAA is determined by the formula:

$$P = \int_{Z_{\text{lim}}}^{\infty} \varphi(Z), \tag{2.11}$$

where $\varphi(Z)$—distribution density function of the pollution indicator; Z_{lim}—limit value of Z for different levels of pollution ($Z_{\text{lim}} = \{Z1, Z2, Z3, Z4\}$).
As indicated above, the distribution of the pollution index of the environmental component (Z) most fully describes by the Weibull distribution law with parameters b and m, depending on the mathematical expectation Z and its coefficient of variation, that is, the variability of Z in an annual or long-term section, or according to different sources of pollution:

$$\varphi(Z) = \varphi(x) = \left(\frac{b}{\overline{x}}\right)^m \times x^{m-1} \times \exp\left[-\left(\frac{bx}{\overline{x}}\right)^m\right] \tag{2.12}$$

where b and m—parameters of the Weibull distribution; $\overline{x}$—average value of the SLAA hazard index of the estimated time interval.
Substituting expression (2.11) instead of $\varphi(Z)$ in formula (2.10) and taking the integral within the indicated limits, it can be obtained:

a. probability of risk of moderate pollution:

$$P_2 = \exp\left[-\left(\frac{Z_1 \times b}{M(Z)}\right)^m\right] = e^{-\left(\frac{Z_1 \times b}{M(Z)}\right)^m}, \tag{2.13}$$

b. probability of risk of dangerous pollution for a sensitive group of the population:

$$P_3 = \exp\left[-\left(\frac{Z_2 \times b}{M(Z)}\right)^m\right] = e^{-\left(\frac{Z_2 \times b}{M(Z)}\right)^m}, \tag{2.14}$$

c. probability of risk of dangerous pollution:

$$P_4 = \exp\left[-\left(\frac{Z_3 \times b}{M(Z)}\right)^m\right] = e^{-\left(\frac{Z_3 \times b}{M(Z)}\right)^m}, \tag{2.15}$$

d. probability of risk of very dangerous pollution:

$$P_5 = \exp\left[-\left(\frac{Z_4 \times b}{M(Z)}\right)^m\right] = e^{-\left(\frac{Z_4 \times b}{M(Z)}\right)^m}, \tag{2.16}$$

where $M(Z)$—mathematical expectation and the average value of Z in the estimated period; b and m—parameters of the Weibull distribution.

Thus, the specific expressions of the obtained dependencies (2.13–2.16) are probable values, that is, they characterize the risk probability depending on the intensity of various sources of SLAA pollution of the UA and the absolute (average) value of the atmospheric air hazard index. For calculating the risk, it is necessary to know only the average value of Z for pollution sources or at a specific time interval and the coefficient of variation of the pollution indicator for these sources or for a given time interval. This criterion, which varies from 0 to 1 (or from 0 to 100%), allows in a convenient form to assess the risks of contamination of environmental components at any UA.

6. Risk assessment is carried out to determine the expected value of losses from exposure of an individual risk stressor and assess whether the consequences are significant enough to require enhanced management (or regulation) in the situation under consideration.

Table 2.7 Risk level classification

Risk level	Lifetime risk
High (De Manifestis)—not acceptable for working conditions and population. It is necessary to implement measures to eliminate or reduce the risk	$>10^{-3}$
Medium—acceptable for production conditions; with an impact on the entire population, dynamic control and an in-depth study of the sources and possible consequences of harmful effects are required to resolve the issue of risk management actions	10^{-3} to 10^{-4}
Low—acceptable risk (the level at which, as a rule, hygiene standards for the population are established	10^{-4} to 10^{-6}
Minimum (De Minimis)—desirable (target) value of the risk during health improvement and environmental protection actions	$<10^{-6}$

Individual risk R_i can be represented by the formula (2.17):

$$R_i = \frac{P_{el} \times P_{ap} \times T}{N} \tag{2.17}$$

where P_{el}—probability of exceeding the permissible hazard limit for certain types of atmospheric air pollution, P_{ap}—probability of the risk of hazardous air pollution for the population, T—period for which the risk needs to be calculated, N—number of population living in the district where the PMP are located.

7. During assessing health risks associated with exposure to atmospheric air, it is advisable to focus on the system of criteria recommended by the World Health Organization (Table 2.7).

2.4 Conclusions

1. The algorithm and stages of the study of the state of the environment of the UAs territory are proposed. Environmental studies were carried out on the territory of the capital of Ukraine—Kyiv. It is shown that during assessing the chemical pollution of SLAA (total), it should take into account the hazard class of pollutants, the frequency of excess of the MPC of the impurity, the nature of the combined action of substances that are jointly present in the air. Recently, a more informative indicator has become widespread in hygienic research—the index of urban air pollution (API).
2. Physical (noise) pollution is an important parameter for assessing the quality of SLAA of UA. Environmental and hygienic regulation is carried out in such a way as to satisfy the listed requirements and principles of environmental legislation and sanitary and hygienic legislation.

3. The main approaches to the assessment of environmental risk under the conditions of technogenic loads have been investigated and the corresponding mathematical models of the statistically evaluated category, which are a vector multicomponent quantity, have been obtained.
4. A method for assessing the hazard of atmospheric air in UA is proposed according to the following algorithm:
   - determination of priority factors of atmospheric air pollution in UA;
   - standardization of the values of the criteria and the determination of the indices of the state of atmospheric air according to individual criteria;
   - calculation of hazard indices of atmospheric air by chemical and physical factors;
   - improving the assessment of the complex hazard of atmospheric air in UA;
   - provision of a graphical result of atmospheric air hazard based on GIS.
5. The concept of assessing environmental risks at the existing level of technogenic impact, when the environment and human health are considered as the object of impact, is proposed. A method for calculating the determination of the individual risk of atmospheric air hazard for the population and the probability of exceeding the atmospheric air quality standards has been developed.

References

1. Zaporozhets, A., Babak, V., Isaienko, V., Babikova, K.: Analysis of the air pollution monitoring system in Ukraine. In: Babak V., Isaienko V., Zaporozhets A. (eds.) Systems, Decision and Control in Energy I. Studies in Systems, Decision and Control, vol. 298, pp. 85–110. Springer, Cham (2020). https://doi.org/10.1007/978-3-030-48583-2_6
2. Sokolovska, A.V.: Space monitoring of ecological condition of urban territories (the Kyiv city is used as an example). Space Sci. Technol. **4**(83), 44–49 (2013). https://doi.org/10.15407/knit2013.04.044
3. Isaenko, V., Zaporozhets, A., Babikova, K., Gulevets, D., Savchenko, S.: Review of methods and means on monitoring the air pollution. Proc. Natl. Aviat. Univ. **3**(80), 61–70 (2019). https://doi.org/10.18372/2306-1472.80.14275
4. Babich, H., Lighthart, B.: Air pollution and microbial ecology. CRC Crit. Rev. Environ. Control. **4**(1–4), 353–421 (1974). https://doi.org/10.1080/10643387409381619
5. Cappitelli, F., Fermo, P., Vecchi, R., Piazzalunga, A., Valli, G., Zanardini, E., Sorlini, C.: Chemical–physical and microbiological measurements for indoor air quality assessment at the Ca' Granda historical archive, Milan (Italy). Water Air Soil Pollut. **201**, 109–120 (2009). https://doi.org/10.1007/s11270-008-9931-5
6. Goudarzi, G., Geravandi, S., Idani, E., et al.: An evaluation of hospital admission respiratory disease attributed to sulfur dioxide ambient concentration in Ahvaz from 2011 through 2013. Environ. Sci. Pollut. Res. **23**, 22001–22007 (2016). https://doi.org/10.1007/s11356-016-7447-x
7. Smith, S.J., van Aardene, J., Klimont, Z., Andres, R.J., Volke, A., Delgado Arias, S.: Anthropogenic sulfur dioxide emissions: 1850–2005. Atmos. Chem. Phys. **11**(3), 1101–1116 (2011). https://doi.org/10.5194/acp-11-1101-2011

8. Boningari, T., Smirniotis, P.G.: Impact of nitrogen oxides on the environment and human health: Mn-based materials for the NO_x abatement. Curr. Opin. Chem. Eng. **13**, 133–141 (2016). https://doi.org/10.1016/j.coche.2016.09.004
9. Peel, J.L., Haeuber, R., Garcia, V., Russel, A.G., Neas, L.: Impact of nitrogen and climate change interactions on ambient air pollution and human health. Biogeochemistry **114**, 121–134 (2013). https://doi.org/10.1007/s10533-012-9782-4
10. Jhun, I., Coull, B.A., Zanobetti, A., Koutrakis, P.: The impact of nitrogen oxides concentration decreases on ozone trends in the USA. Air Qual. Atmos. Health **8**, 283–292 (2015). https://doi.org/10.1007/s11869-014-0279-2
11. Logan, J.A.: Nitrogen oxides in the troposphere: global and regional budgets. J. Geophys. Res. Atmos. **88**(C15), 10785–10807 (1983). https://doi.org/10.1029/JC088iC15p10785
12. Logan, J.A.: Tropospheric ozone: seasonal behavior, trends, and anthropogenic influence. J. Geophys. Res. Atmos. **90**(D6), 10463–10482 (1985). https://doi.org/10.1029/JD090iD06p10463
13. Liu, H., Jacob, D.J., Chan, L.Y., Oltmans, S.J., Bey, I., Yantosca, R.M., Harris, J.M., Duncan, B.N., Martin, R.V.: Sources of tropospheric ozone along the Asian Pacific Rim: an analysis of ozonesonde observations. J. Geophys. Res. Atmos. **107**(D21), ACH 3-1–ACH 3-19 (2002). https://doi.org/10.1029/2001JD002005
14. So, K.L., Wang, T.: On the local and regional influence on ground-level ozone concentrations in Hong Kong. Environ. Pollut. **123**(2), 307–317 (2003). https://doi.org/10.1016/S0269-7491(02)00370-6
15. Ilgen, E., Karfich, N., Levsen, K., Angerer, J., Schneider, P., Heinrich, J., Wichmann, H.-E., Dunemann, L., Begerow, J.: Aromatic hydrocarbons in the atmospheric environment: Part I. Indoor versus outdoor sources, the influence of traffic. Atmos. Environ. **35**(7), 1235–1252 (2001). https://doi.org/10.1016/S1352-2310(00)00388-5
16. Jobson, B.T., Berkowitz, C.M., Kuster, W.C., Goldan, P.D., Williams, E.J., Fesenfeld, F.C., Apel, E.C., Karl, T., Lonneman, W.A., Riemer, D.: Hydrocarbon source signatures in Houston, Texas: influence of the petrochemical industry. J. Geophys. Res. Atmos. **109**(D24) (2004). https://doi.org/10.1029/2004JD004887
17. Lu, H., Zhu, L., Zhu, N.: Polycyclic aromatic hydrocarbon emission from straw burning and the influence of combustion parameters. Atmos. Environ. **43**(4), 978–983 (2009). https://doi.org/10.1016/j.atmosenv.2008.10.022
18. Xu, L., Jiang, X.: Notice of retraction: control of exhaust pollutants from gasoline engine of vehicle. In: 2010 International Conference on Computer Application and System Modeling (ICCASM 2010), pp. V13-564–V13-567. Taiyuan (2010). https://doi.org/10.1109/iccasm.2010.5622903
19. Zhdanko, I.M., Zinkin, V.N., Soldatov, S.K., Bogomolov, A.V., Sheshegov, P.M.: Fundamental and applied aspects of preventing the adverse effects of aviation noise. Human Physiol. **42**, 705–714 (2016). https://doi.org/10.1134/S0362119716070227
20. Petrivskyi, V., Shevchenko, V., Bychkov, O., Brazhenenko, M.: Estimation of noise hazards in environmental monitoring tools designed in the subway. In: 2019 IEEE 15th International Conference on the Experience of Designing and Application of CAD Systems (CADSM), pp. 1–4. Polyana, Ukraine (2019). https://doi.org/10.1109/cadsm.2019.8779315
21. Berezutskyi, V., Illinska, O., Vaskovets, L., Mezentseva, I.: Determining the effect of multifunctional partitions on noise level at open space offices. Eastern-European J. Enterp. Technol. **2/10**(98), 16–26 (2019). https://doi.org/10.15587/1729-4061.2019.164786
22. Burichenko, L.A., Gulevets, V.D.: Safety in Aviation: A Textbook. National Aviation University, Kyiv (2003). (in Ukrainian)
23. Faisal, A.A., Selen, L.P.J., Wolpert, D.M.: Noise in the nervous system. Nat. Rev. Neurosci. **9**, 292–303 (2008). https://doi.org/10.1038/nrn2258
24. Vehlow, J.: Air pollution control systems in WtE units: An overview. Waste Manag. **37**, 58–74 (2015). https://doi.org/10.1016/j.wasman.2014.05.025

25. Mosley, S.: Environmental history of air pollution and protection. In: Agnoletti M., Neri Serneri S. (eds.) The Basic Environmental History. Environmental History, vol. 4, pp. 143–169. Springer, Cham (2014). https://doi.org/10.1007/978-3-319-09180-8_5
26. Ancillotti, E., Bruno, R., Conti, M.: The role of communication systems in smart grids: architectures, technical solutions and research challenges. Comput. Commun. **36**(17–18), 1665–1697 (2013). https://doi.org/10.1016/j.comcom.2013.09.004
27. Iatsyshyn, A, Iatsyshyn, A, Artemchuk, V., Kameneva, I., Kovach, V., Popov, O.: Software tools for tasks of sustainable development of environmental problems: peculiarities of programming and implementation in the specialists' preparation. In: E3S Web of Conferences 166, 01001 (2020). https://doi.org/10.1051/e3sconf/202016601001
28. Zhukov, O.V., Potapenko, O.V.: Environmental impact assessment of distribution substations: The case of phytoindication. Ukrainian J. Ecol. **7**(1), 5–21 (2017). (in Ukrainian)
29. Lee, D., Rushworth, A., Sahu, S.K.: A Bayesian localized conditional autoregressive model for estimating the health effects of air pollution. Biometrics **70**(2), 419–429 (2014). https://doi.org/10.1111/biom.12156
30. Qi, Z., Wang, T., Song, G., Hu, W., Li, X., Zhang, Z.: Deep air learning: interpolation, prediction, and feature analysis of fine-grained air quality. IEEE Trans. Knowl. Data Eng. **30** (12), 2285–2297 (2018). https://doi.org/10.1109/tkde.2018.2823740
31. Yuval, Levy, I., Broday, D.M.: Improving modeled air pollution concentration maps by residual interpolation. Sci. Total Environ. **598**, 780–788 (2017). https://doi.org/10.1016/j.scitotenv.2017.04.117
32. Chizari, H., Hosseini, M., Poston, T., Abd Razak, S., Abdullah, A.H.: Delaunay triangulation as a new coverage measurement method in wireless sensor network. Sensors **11**(3), 3163–3176 (2011). https://doi.org/10.3390/s110303163
33. Chen, M.-B.: A Parallel 3D delaunay triangulation method. In: 2011 IEEE Ninth International Symposium on Parallel and Distributed Processing with Applications, pp. 52–56. Busan (2011). https://doi.org/10.1109/ispa.2011.52
34. Faber, J.H., Marshall, S., Van den Brink, P., Maltby, L.: Priorities and opportunities in the application of the ecosystem services concept in risk assessment for chemicals in the environment. Sci. Total Environ. **651**(1), 1067–1077 (2019). https://doi.org/10.1016/j.scitotenv.2018.09.209
35. Shi, L., Zhang, M., Yang, B., Gao, L.: Air pollution-oriented ecological risk assessment in Xiamen city, China. Int. J. Sustain. Dev. World Ecol. **25**(5), 420–430 (2018). https://doi.org/10.1080/13504509.2017.1419390
36. Popov, O.O., Iatsyshyn, A.V., Kovach, V.O., Artemchuk, V.O., Kameneva, I.P., Taraduda, D.V., Sobyna, V.O., Sokolov, D.L., Dement, M.O., Yatsyshyn, T.M.: Risk assessment for the population of Kyiv, Ukraine as a result of atmospheric air pollution. J. Health Pollut. **10**(25), 200303 (2020). https://doi.org/10.5696/2156-9614-10.25.200303

Chapter 3
Research of Chemical and Physical Pollution in Kyiv City

Abstract In this chapter, a study of the complex influence of physical and chemical factors on the state of the environment is carried out using the example of the Kyiv city. To carry out research on the chemical pollution of the surface layer of the atmospheric air of the Kyiv city, a database was formed according to the data of the Central Geophysical Observatory. Statistical samples of pollutant concentrations are characterized by large volumes and include a complete list of pollutants that were monitored. Full-scale acoustic measurements were carried out in the vicinity of the international airport "Kyiv". For the current traffic intensity, contour areas of equivalent sound levels were calculated for day and night.

Keywords Acoustic measurements · Pollutants · Statistical analysis · Monitoring · Kyiv international airport

3.1 General Characteristics of Kyiv City and the Object of Full-Scale Measurements

Modern Kyiv city is one of the ten largest cities in Europe. The total area of the Kyiv city is 835.6 km^2. The built-up land of the city is 364.0 km^2 or 43.5%, of which 115.0 km^2 are under residential and public buildings. A significant amount of land is occupied by industrial facilities—56.0 km^2, transport and communication facilities—22.0 km^2.

The originality and diversity of the natural conditions of Kyiv are associated with its location on the border of the physical and geographical zones: the forest-steppe zone and the zone of mixed forests. The northern part of the city is located on the Polissya lowland, the southwestern (right bank)—on the Dnipro upland, the southeastern (left bank)—on the Dnipro lowland.

The climate of Kyiv is moderately continental, with rather mild winters and warm summers. A tangible influence on the climate of Kyiv is carried out by the Dnipro river, which is stretched in the submeridional direction within the city. A large movable water area contributes to the formation of air transport breezes:

D. Gulevets et al., *Environmental Control for Ensuring Cities Safety*,
Lecture Notes in Intelligent Transportation and Infrastructure,
https://doi.org/10.1007/978-3-030-66710-8_3

during the day the temperature difference between the water surface and land creates streams of fresh moist air into the city. During the year, anticyclonic activity prevails, which is characterized by rather stable, slightly cloudy weather. The average annual air temperature is +8.9…+11.9 °C. Global climate changes, observed on the globe, could not bypass the Kyiv city. Moreover, the climatic conditions are significantly influenced by the city itself—the dissipation of heat from heating pipes, buildings, thermal power plants, etc. In this regard, the air temperature in the city is higher than in its vicinity. The increase in air temperature of Kyiv over the past decades is greater than the global one on the planet. The average annual rainfall is 600–700 mm [1, 2].

The residential area is characterized by high-rise buildings in the central, right-bank part of the city, on the massifs—Obolon, Vinogradar, Teremky, on the Left Bank—Troeshchina and Kharkiv massifs. Private buildings are located mainly in the vicinity of the city along its perimeter.

The industrial zone consists of industrial and road transport enterprises. Within the boundaries of the Kyiv urban agglomeration, they are grouped into industrial centers and zones: Podilsko-Obolonsky, Shulyavka, Nizhnyolibidsky, Darnitsky, Telbinsky.

The recreational zone is represented by conventionally natural landscapes (forests, meadows, lakes, rivers), which have been preserved within the city and its environs by artificial green spaces. It has a positive effect on the environment and is an indicator of ecological well-being.

The technique for measuring noise pollution on the territory of residential development is regulated by [3]. There are regulatory requirements to obtain noise characteristics by full-scale measurements of individual noise sources.

IA “Kyiv” is located in the Solomensky district of the city. The airport has one runway with artificial turf—2310 × 45 m. The effect of noise generated by aircraft arriving or departing from the airport depends on a number of factors, namely:

- types of aircraft operated at the airport aerodrome;
- number of aircraft takeoffs and landings in the busiest week (determined according to the airport data);
- distribution of aircraft by routes (determined by airport data);
- distribution of aircraft by the period of the day (determined according to the airport data);
- number of runways and their use;
- weather conditions and flight rules specific to the airport.

The process of investigating the effects of noise around airports involves modeling “noise contours”. The noise contour is a line of constant value for the sound level of aircraft noise, averaged over a period of time, that is generated during the passage of aircraft of various types under normal operating conditions. The permissible levels and requirements for buildings in terms of the influence of aircraft noise are determined in accordance with [4].

The standard parameters define the equivalent sound level and the maximum sound level. The regulated intervals for monitoring the impact of aircraft noise are: the daytime period (from 7.00 to 23.00—16 h) and the night-time period (from 23.00 to 7.00—8 h) [5, 6].

Tables 3.1 and 3.2 show the noise levels and areas of possible development around the airport. Zone D defines an unsuitable territory for residential development, zone B and zone C—territories with restrictions and requirements for the application of actions to reduce noise impact, and zone A is a territory without restrictions for the building's construction of any purpose.

The boundaries of the zones of protection against the adverse effects of aircraft noise and the suitability of the territories around airports and aerodromes for residential and public buildings are determined using the calculation method for assessing noise levels LA_{eq} and LA_{max} [7, 8] and the computer program (INM Pre-approved List of Aircraft Substitutions) for construction of noise contours for standard values of sound levels and plotting contours on a map with reference to the RS.

Noise contours that meet regulatory targets define the boundaries of noise zones around airports. Full-scale acoustic measurements in the vicinity of Kyiv airport were carried out according to [3].

The following instruments were used during the measurements:

- portable noise monitoring station SVAN 958 No. 23436 with GRASS 40AE No. 178052 (certificate of verification of working measuring instruments No. 22-02/1309967);
- sound level meter SVAN 979 No. 34063 with GRASS 40AE No. 178254 (certificate of verification of working measuring equipment No. 22-10/1309972);
- sound level meter SVAN 979 No. 34064 with GRASS 40AE No. 178263 (certificate of verification of working measuring instruments No. 22-10/1309971).

The measurements were carried out in the vicinity of the Kyiv airport at the following noise observation points (NMPs), namely:

- measurement point No. 1a 50°24′20.83″ N, 30°29′31.86″ E, height above sea level—151 m, distance from the end 26—1.8 km, distance from the glide path —180 m;

Table 3.1 Zones that determine the suitability of the territory in the vicinity of the airport for development

Time period	A	B	C	D
Day	$LA_{eq} \leq 60$ $LA_{max} \leq 80$	$61 \leq LA_{eq} \leq 65$ $81 \leq LA_{max} \leq 85$	$61 \leq LA_{eq} \leq 65$ $81 \leq LA_{max} \leq 85$	$LA_{eq} > 65$ $LA_{max} > 85$
Night	$LA_{eq} \leq 50$ $LA_{max} \leq 70$	$51 \leq LA_{eq} \leq 55$ $71 \leq LA_{max} \leq 75$	$56 \leq LA_{eq} \leq 60$ $76 \leq LA_{max} \leq 80$	$LA_{eq} > 60$ $LA_{max} > 80$

Table 3.2 Possibility of building's construction in restricted areas around the airport

Building's purpose	Building's construction in zones			
	A	B	C	D
Residential buildings, kindergartens	Allowed	Allowed with increased sound insulation of external fences, providing noise reduction, ΔLA dBA		Prohibited
		25	30	
Polyclinics	Allowed in the part of the zone with levels in the daytime $LA_{eq} \leq$ 55 dBA without limitation, for $LA_{eq} \leq$ 56–60 dBA with increased sound insulation (ΔLA = 25 dBA)	Allowed with increased sound insulation, which provides ΔLA = 30 dBA		Prohibited
Schools and other educational institutions	Allowed	Allowed with increased sound insulation, which provides ΔLA = 25 dBA		Prohibited
Hotels, hostels	Allowed	Allowed with increased sound insulation, which provides ΔLA, dBA		Prohibited
		25	30	
Administrative buildings, design and research organizations	Allowed	Allowed	Allowed	Allowed with the necessary sound insulation

- measurement point No. 1b 50°24′16.24″ N, 30°29′3.40″ E, height above sea level—147 m, distance from the end 26—1.23 km, distance from the glide path—60 m;
- measurement point No. 2 50°24′7.61″ N, 30°28′38.50″ E, height above sea level—173 m, distance from the end 26—0.72 km, distance from the glide path—95 m;
- measurement point No. 3 50°24′3.34″ N, 30°25′5.92″ E, height above sea level—168 m, distance from end face 08—1.17 km, distance from the glide path—50 m;
- measurement point No. 4 50°23′58.97″ N, 30°24′1.37″ E, height above sea level—171 m, distance from end face 08—2.44 km, distance from glide path—50 m (Figs. 3.1 and 3.2).

Fig. 3.1 Placement of devices in relation to the IA "Kyiv"

3.2 Calculation of the Hazard Index of SLAA from Chemical Pollution

The monitoring of the state of pollution of the SLAA in Kyiv was carried out by the CGO at 16 stationary posts on the content of the main pollutants, which include dust, sulfur dioxide, carbon monoxide, nitrogen dioxide, and specific substances (nitric oxide, dissolved sulfates, hydrogen sulfide, phenol, hydrogen fluoride, hydrogen chloride, ammonia, formaldehyde, benzopyrene, heavy metals) (Fig. 3.3). The numbers at the marks indicate the number of the PMP or field measurements:

- № 1: 6A, Strazhesko Str.;
- № 2: 8, Dovzhenko Str. (Shulyavska);
- № 3: 50, Popudrenko Str. (Chernihivska);
- № 4: 2, Lazo Str.;
- № 5: 37, Nauki Ave.;
- № 6: Victory Square;
- № 7: Bessarabian Square;
- № 8: 29, Lesia Ukrainka Boulevard;
- № 9: 10A, Kaunaska Str.;
- № 10: 56/60, Mezhyhirska Str. (Shevchenko);
- № 11: 98/2, Victory Avenue (Svyatoshin);
- № 13: Expocenter of Ukraine;
- № 15: Hydropark;
- № 17: 14, Obolonsky Ave. (Obolon);
- № 20: Demiivska Square;
- № 21: 25, Sklyarenka Str.

Fig. 3.2 Photos of experimental research

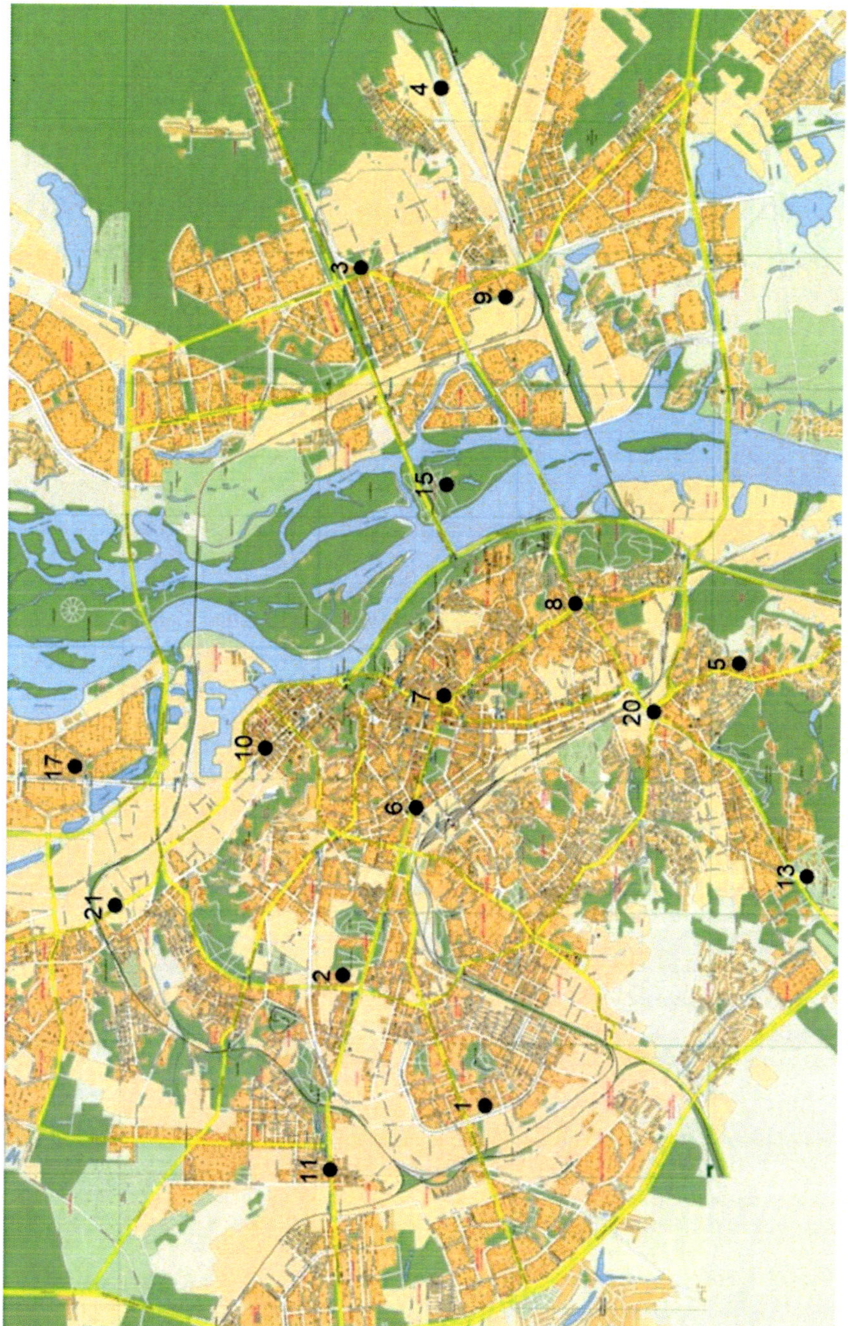

Fig. 3.3 Location of sampling points to assess the content of pollutants in the SLAA of the Kyiv city

For carrying out research on chemical pollution, a database has been formed based on the data of the CGO. Statistical samples of pollutant concentrations are characterized by high volumes, which cover 2010–2013 and include a complete list of pollutants [9, 10].

According to the SLAA hazard index, the general level of pollution in the whole city in 2010–2013 characterized as a dangerous level for a sensitive group of the population.

The highest pollution indices of SLAA were noted on Bessarabian Square (PMP № 7), Victory Square (PMP № 6), Oleksandr Dovzhenko Street (PMP № 2), Obolonsky Avenue (PMP № 17) and Akademika Strazhesko Street (PMP № 1) (Fig. 3.4).

The highest average annual SLAA hazard indices were noted on Bessarabian Square, Victory Square, Victory Avenue and Sklyarenko Str.—231, 221, 212 and 205 respectively, which is characterized as a dangerous level for a sensitive group of the population (Table 3.3). The average annual indices of two pollutants exceeded the safety limit: formaldehyde (substance of the second hazard class)—by 2.3 times, nitrogen dioxide (substance of the third hazard class)—by 3.5 times.

Table 3.3 shows the calculations of the indices and the average value for each PMP in the period from 2010–2013.

According to calculations, the largest excess on average was: nitrogen dioxide—47.8%, formaldehyde—34.5%, suspended solids—8.96%, phenol—6.63%, carbon monoxide—3.37%, nitric oxide—2.15% of the total number of observations in the city (Fig. 3.5).

The level of nitrogen dioxide according to the average annual hazard indices at almost all posts exceeded the permissible level of pollution and averaged: 2010—

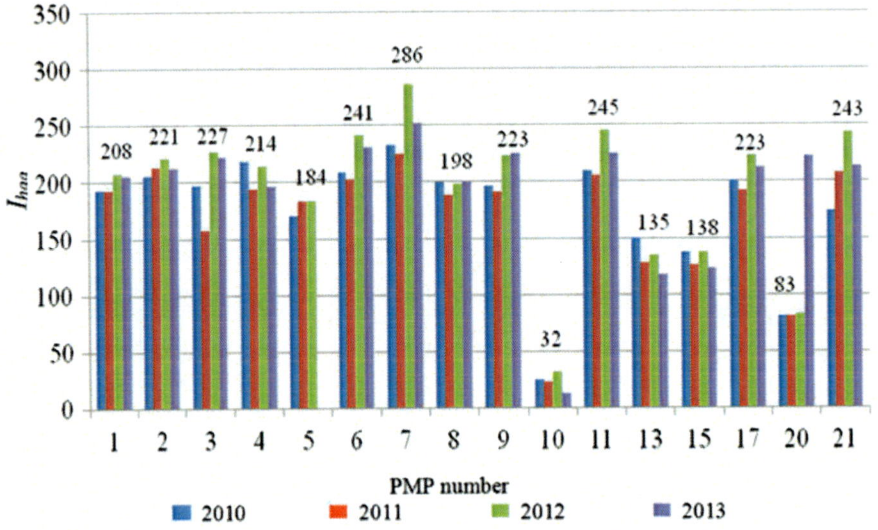

Fig. 3.4 I_{haa} value on PMPs in 2010–2013

Table 3.3 Values of the hazard indices of the SAA (I_{haa}) for the Kyiv city according to the PMP data for 2010–2013

2010		2011		2012		2013		Mean
PMP №	I_{haa}	PMP №	I_{haa}	PMP №	I_{haa}	PMP №	I_{haa}	I_{haa}
7	233	7	225	7	286	7	252	249
4	218	2	213	11	245	6	231	227
11	210	21	208	21	243	9	225	222
6	209	11	206	6	241	11	225	220
2	206	6	203	3	227	3	222	215
17	201	4	194	9	223	20	222	210
8	200	1	193	17	223	21	213	207
3	197	17	192	2	221	2	212	206
9	196	9	191	4	214	17	212	203
1	193	8	188	1	208	1	206	199
21	174	5	186	8	198	8	200	190
5	170	3	158	5	184	4	196	177
13	150	13	128	15	138	15	123	135
15	138	15	126	13	135	13	117	129

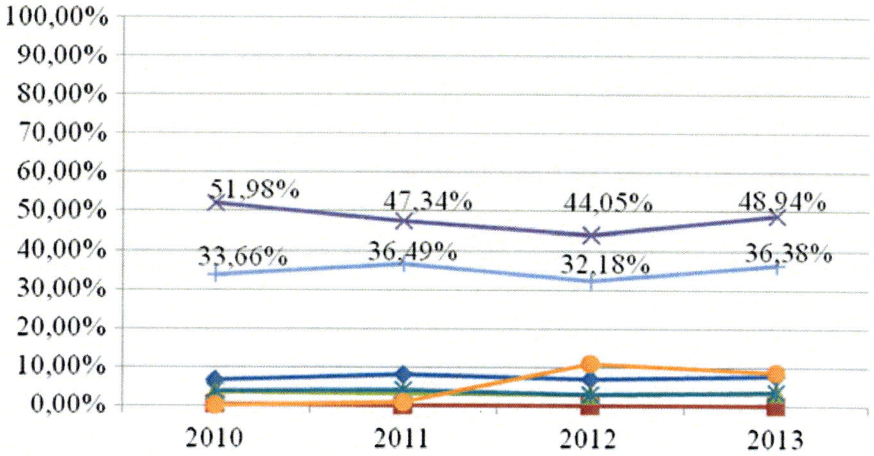

Fig. 3.5 I_{haa} value on PMPs in 2010–2013 in percentage terms

179, 2011—173, 2012—204, 2013—197, which according to the SLAA hazard scale corresponds to a high level of pollution (Fig. 3.6).

The largest number of cases of excess nitrogen dioxide from 2010–2013 was observed in the summer months and on average in the Kyiv city it was 9.2% (Fig. 3.7).

The level of formaldehyde, a substance of the second hazard class, according to the average annual hazard indices, at almost all PMPs exceeded the permissible

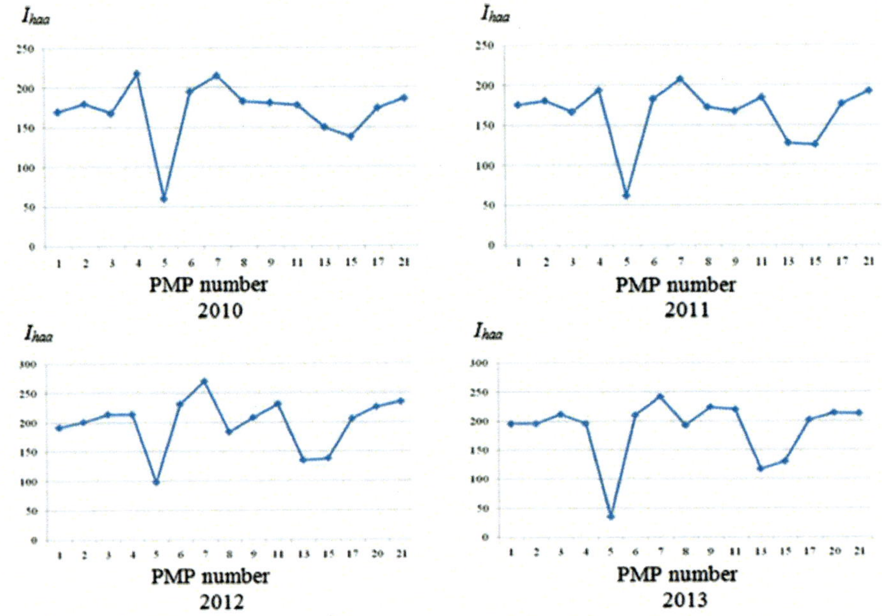

Fig. 3.6 Hazard indices of the SLAA from nitrogen dioxide at the PMPs by 2010–2013 years

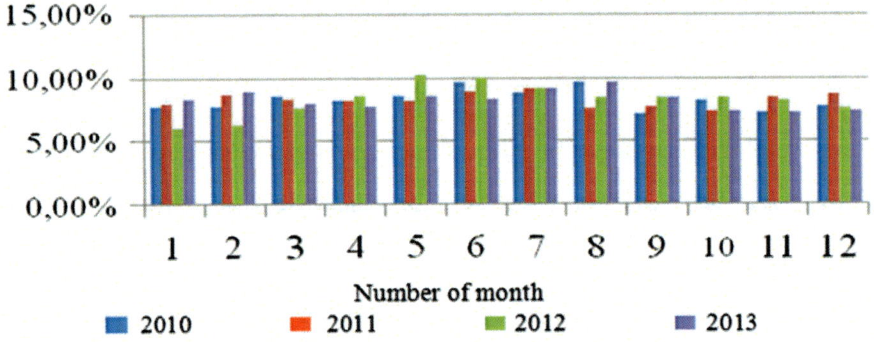

Fig. 3.7 Number of cases of nitrogen dioxide excess from 2010–2013

level of pollution and averaged: 2010—189, 2011—174, 2012—194, 2013—186, which on the SLAA hazard scale corresponds to a high level of pollution (Fig. 3.8).

The largest number of cases of excess of formaldehyde in 2010–2013 was observed most in the summer months and on average in the Kyiv city it was 11.8% (Fig. 3.9).

The level of suspended solids (dust), according to the average annual hazard indices at almost all posts, exceeded the permissible level of pollution and averaged: 2010—106, 2011—103, 2012—101, 2013—114, which on the SLAA hazard scale corresponds to the average level of pollution (Fig. 3.10).

Fig. 3.8 Hazard indices of the SLAA from formaldehyde at the PMPs by 2010–2013 years

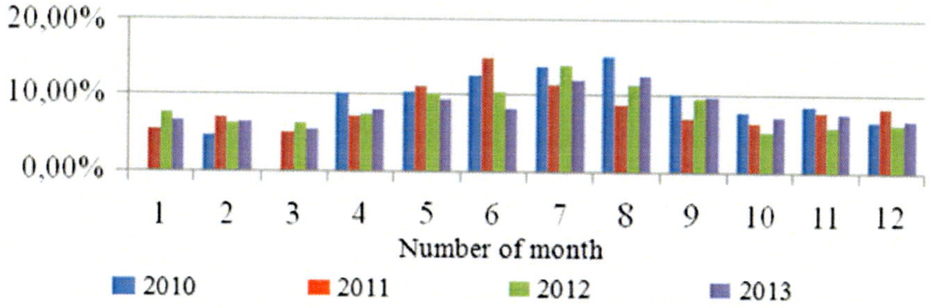

Fig. 3.9 Number of cases of formaldehyde excess from 2010–2013

The largest number of cases of excess of suspended solids (dust) in 2010–2013 was observed almost at the same level and on average in the Kyiv city it was 7.8% (Fig. 3.11).

The level of nitrogen oxide, according to the average annual hazard indices at almost all PMPs, exceeded the permissible level of pollution and averaged: 2010—106, 2011—103, 2012—101, 2013—114, which on the SLAA hazard scale corresponds to the average level of pollution (Fig. 3.12).

The largest number of cases of nitrogen oxide excess in 2010–2013 was observed almost at the same level, only in 2013 there was an excess in the summer months. On average in the Kyiv city it was 7.8% (Fig. 3.13).

The level of carbon monoxide according to the average annual hazard indices in the Bessarabian Square exceeded the permissible level of pollution and averaged:

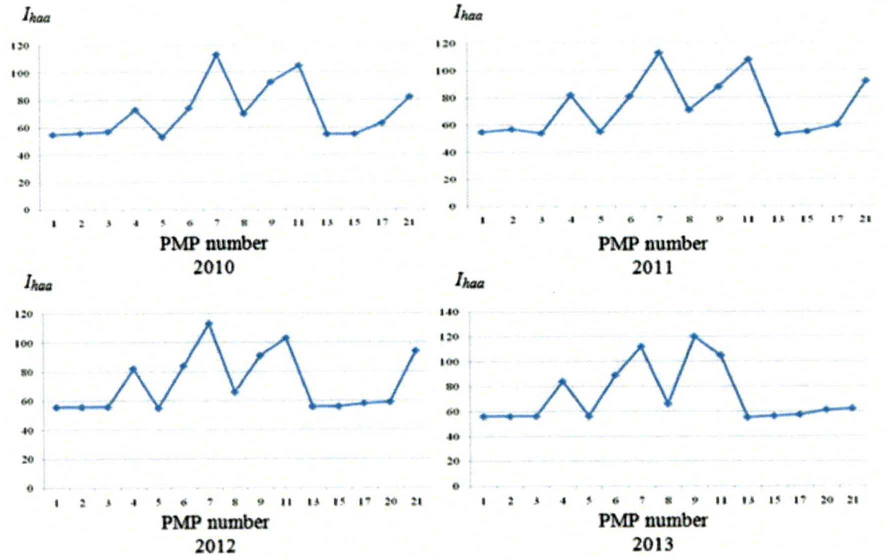

Fig. 3.10 Hazard indices of the SLAA from suspended solids (dust) at the PMPs by 2010–2013 years

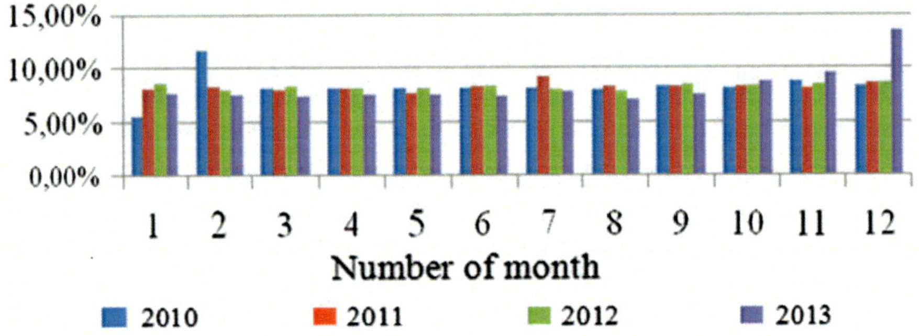

Fig. 3.11 Number of cases of suspended solids (dust) excess from 2010–2013

2010—58, 2011—92, 2012—63, 2013—87, which corresponds to the average level of pollution on the SLAA hazard scale (Fig. 3.14).

The largest number of cases of excess of carbon monoxide in 2010–2013 was observed at almost the same level. On average in the Kyiv city it was 10.8% (Fig. 3.15).

The level of phenol according to the average annual hazard indices exceeded the permissible level of pollution and averaged: 2010—100, 2011—117, 2012—140, 2013—150, which, according to the SLAA hazard scale, corresponds to the average level of pollution (Fig. 3.16).

The largest number of cases of phenol excess in 2010–2013 was observed almost at the same level. On average in the Kyiv city it was 10.8% (Fig. 3.17).

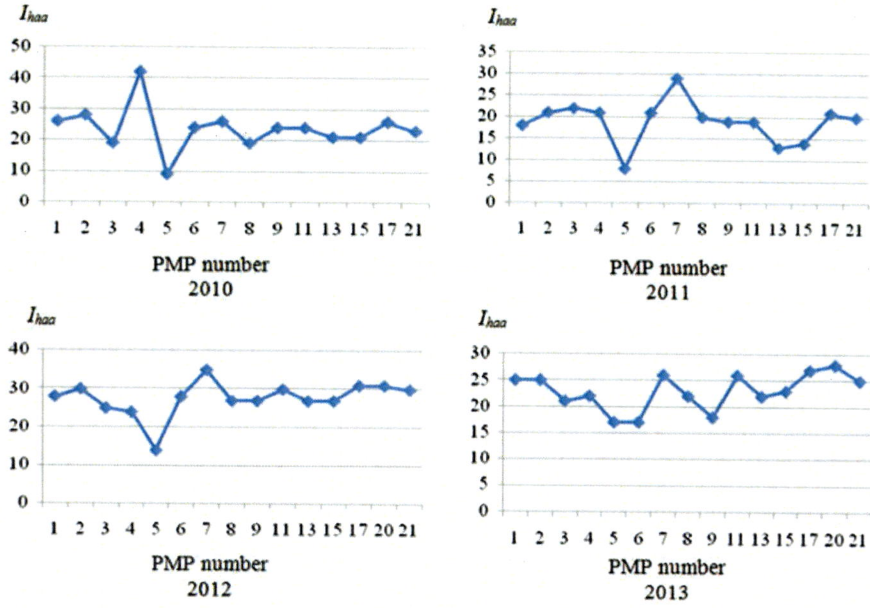

Fig. 3.12 Hazard indices of the SLAA from nitrogen oxide at the PMPs by 2010–2013 years

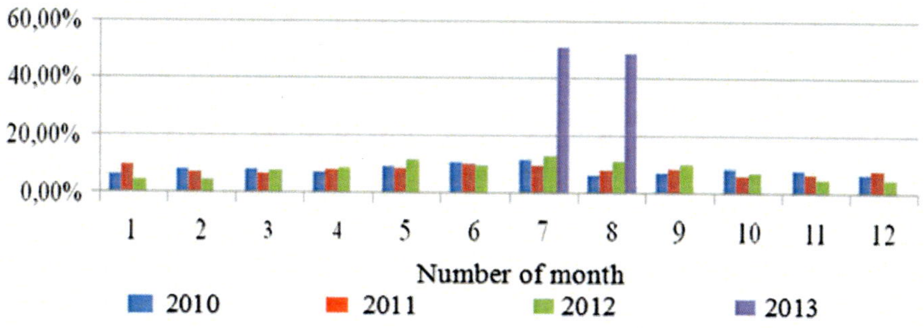

Fig. 3.13 Number of cases of nitrogen oxide excess from 2010–2013

3.3 Calculation of the Hazard Index of the Surface Area of Atmospheric Air from Physical Pollution

The main types of aircraft operated at the airport today: A-320, B-737(s), ATR-75, H25B CL30, CRJ2, PRM1, YK40—on medium highways (Figs. 3.18, 3.19 and 3.20)

Basic principles of calculation:

– Noise exposure levels are determined using the criteria used in national and international hygiene regulations. In Ukraine, noise levels are used as standard

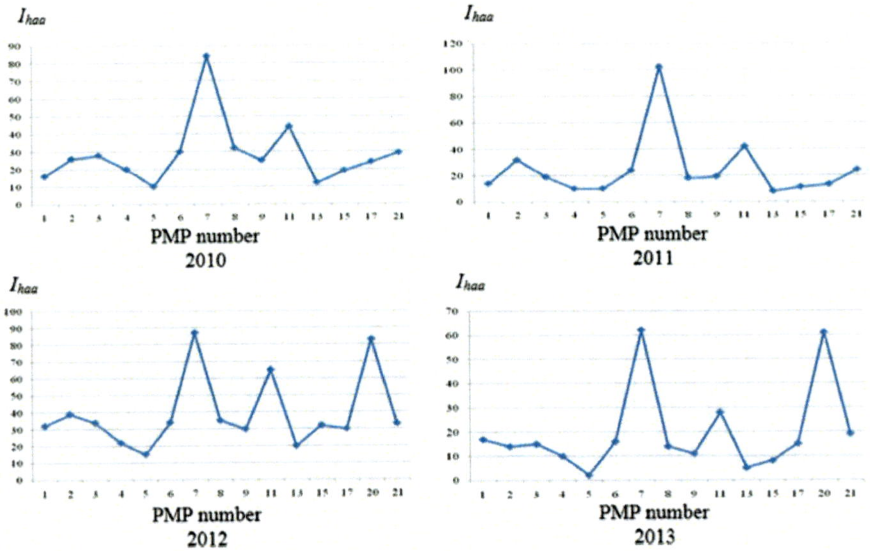

Fig. 3.14 Hazard indices of the SLAA from carbon monoxide at the PMPs by 2010–2013 years

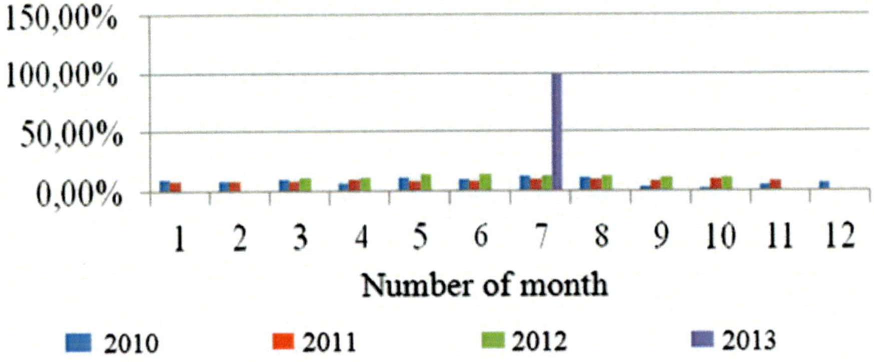

Fig. 3.15 Number of cases of carbon monoxide excess from 2010–2013

values: LA_{max}—maximum noise level with an adjustable frequency band according to the “A” scale of a standard noise meter; LA_{eq}—equivalent noise level. The list of LA_{max} and LA_{eq} values for other noise exposure criteria are performed using the mathematical relationships defined for them.

- Noise levels are determined at the nodes of the computational coordinate grid, which is applied to the surface under investigation at the outskirts of the airport with a fixed step between the nodes. The center of the graticule is located at the airport reference point. The abscissa axis of the coordinate system is directed along the course of movement of aircraft in the airport area.

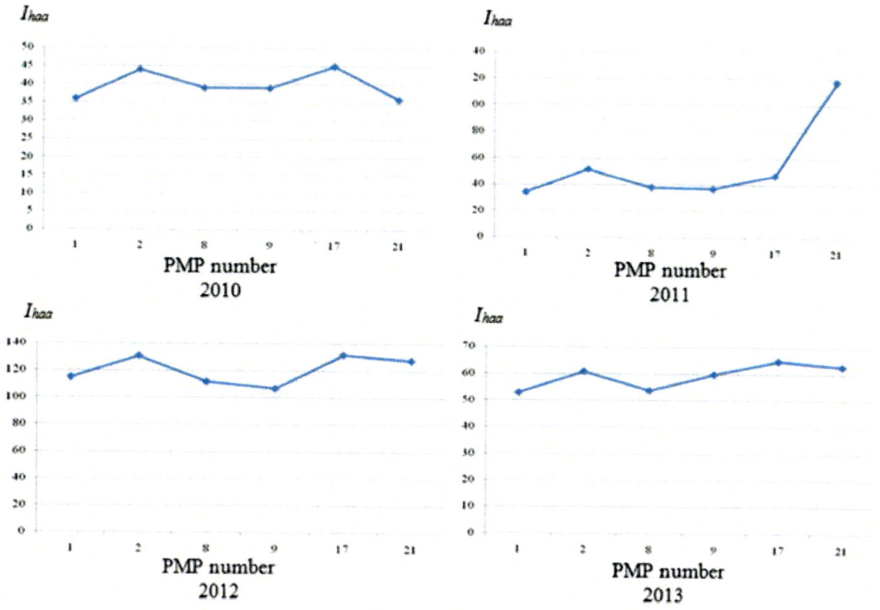

Fig. 3.16 Hazard indices of the SLAA from phenol at the PMPs by 2010–2013 years

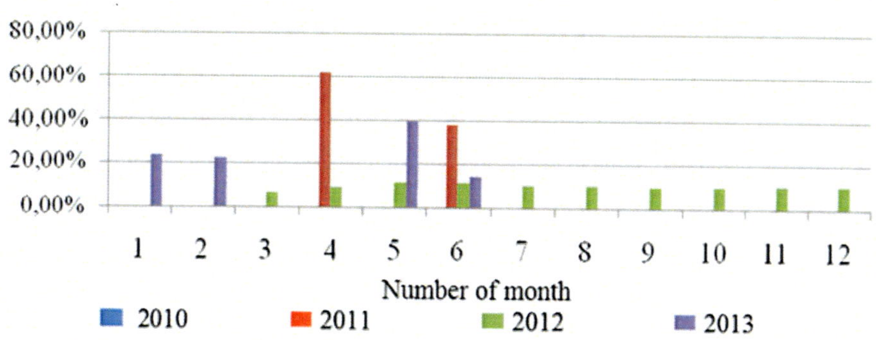

Fig. 3.17 Number of cases of phenol excess from 2010–2013

- The basis for the calculation is the acoustic characteristics of aircraft by type in the form of dependencies: noise level—engine operation mode—distance. The dependence is given based on the results of field studies carried out in accordance with the requirements set before certification studies of aircraft for compliance with the noise requirements on the ground.
- During calculating noise levels, it is necessary to take into account: the real intensity of aircraft traffic in the airport area; distribution of traffic intensity by types of aircraft and traffic routes in the airport area; the availability of noise abatement facilities at the airport (for example, at engine test sites), the use of on-site aircraft noise abatement operational methods or other piloting methods that differ significantly from the nominal ones.

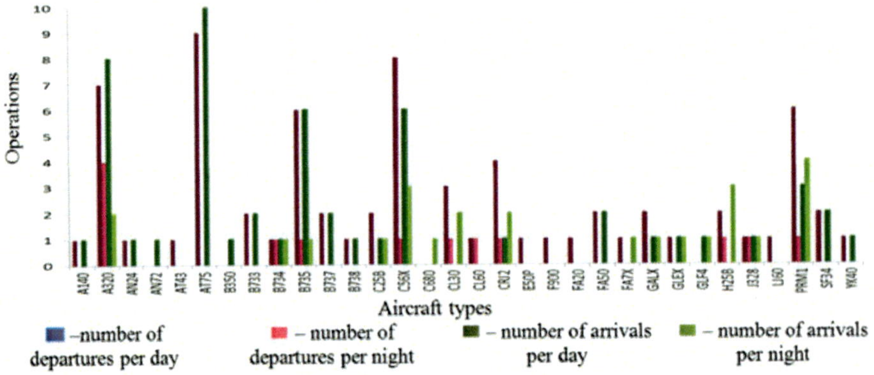

Fig. 3.18 Aircraft fleet structure, busiest day, October 2, 2013

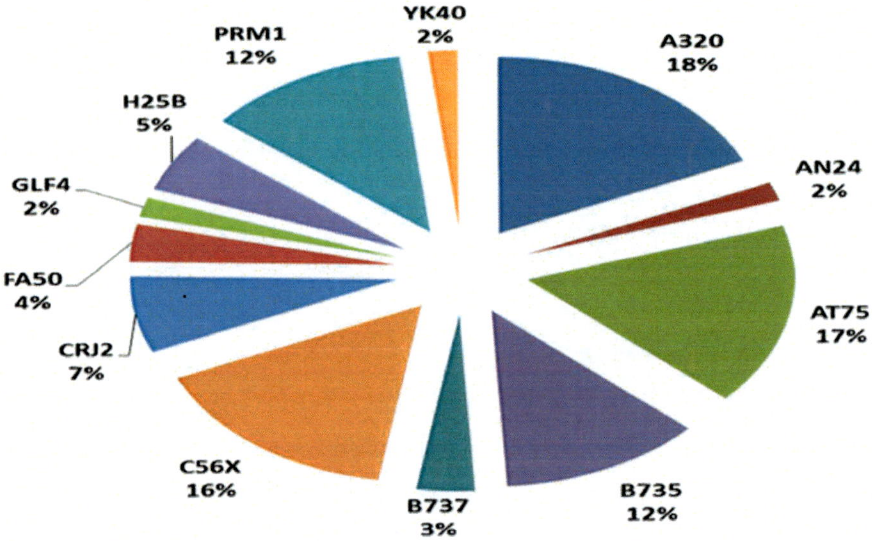

Fig. 3.19 Park structure with a 3% barrier

Noise contours, together with RS, are mapped to the airport location (aerodrome or RS). According to the calculated standard sound levels and the constructed noise contours, the zones of restriction of residential and public administrative buildings are determined in accordance with [4], which completes the calculations.

Taking into account the results of the analysis of the adverse effects of noise for residential and public buildings, the equivalent sound level LA_{eq} and the maximum sound level LA_{max} of 60 dBA and 80 dBA, respectively, in the daytime and 50 dBA and 70 dBA, respectively, at night are acceptable values. The measurement data of noise levels in the vicinity of the Kyiv International Airport are given in the Tables 3.4 and 3.5.

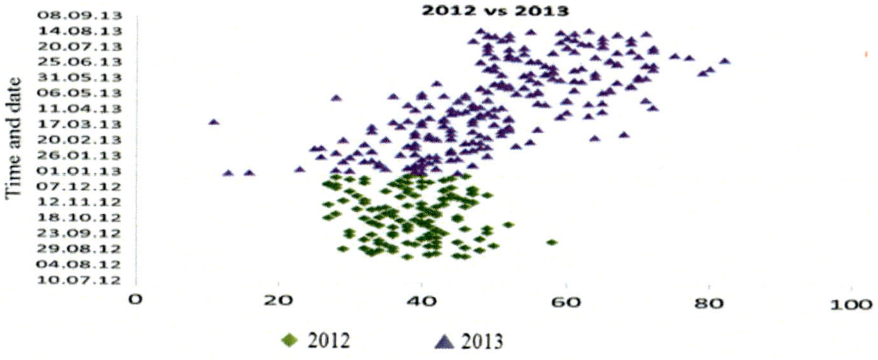

Fig. 3.20 Comparison of the number of take-offs/landings for 2012 and 2013

Table 3.4 Sound levels L_{AE} [dBA] at measurement points for daytime

№	Measurement point 1		Measurement point 2		Measurement point 3		Measurement point 4	
	L_{AE}	$10^{(0.1 \cdot L_{AE})}$	L_{AE}	$10^{(0.1 \cdot L_{AE})}$	L_{AE}	$10^{(0.1 \cdot L_{AE})}$	L_{AE}	$10^{(0.1 \cdot L_{AE})}$
1	96.4	4.37×10^9	100.9	1.23×10^{10}	94.8	3.02×10^9	90.6	1.15×10^9
2	96.3	4.27×10^9	98.6	7.24×10^9	94.5	2.82×10^9	90.5	1.12×10^9
3	96.0	3.98×10^9	96.7	4.68×10^9	94.2	2.63×10^9	90.0	1.00×10^9
4	95.2	3.31×10^9	96.1	4.07×10^9	94.1	2.57×10^9	90.0	1.00×10^9
5	94.6	2.88×10^9	95.9	3.89×10^9	93.9	2.45×10^9	88.9	7.76×10^8
6	94.0	2.51×10^9	95.9	3.89×10^9	93.3	2.14×10^9	88.8	7.59×10^8
7	92.8	1.91×10^9	95.8	3.80×10^9	93.1	2.04×10^9	88.7	7.41×10^8
8	92.7	1.86×10^9	95.4	3.47×10^9	92.9	1.95×10^9	88.3	6.76×10^8
9	92.5	1.78×10^9	95.4	3.47×10^9	92.6	1.82×10^9	88.3	6.76×10^8
10	92.0	1.58×10^9	94.4	2.75×10^9	92.5	1.78×10^9	88.2	6.61×10^8
11	91.1	1.29×10^9	94.2	2.63×10^9	92.5	1.78×10^9	88.2	6.61×10^8
12	90.9	1.23×10^9	94.1	2.57×10^9	92.4	1.74×10^9	88.1	6.46×10^8
13	89.9	9.77×10^8	94.0	2.51×10^9	92.3	1.70×10^9	88.1	6.46×10^8
14	89.3	8.51×10^8	93.5	2.24×10^9	92.1	1.62×10^9	88.0	6.31×10^8
15	89.1	8.13×10^8	93.5	2.24×10^9	92.0	1.58×10^9	87.7	5.89×10^8
16	89.0	7.94×10^8	93.3	2.14×10^9	91.9	1.55×10^9	87.7	5.89×10^8
17	88.9	7.76×10^8	93.3	2.14×10^9	91.8	1.51×10^9	87.3	5.37×10^8
18	88.8	7.59×10^8	93.1	2.04×10^9	91.5	1.41×10^9	86.8	4.79×10^8
19	88.7	7.41×10^8	93.0	2.00×10^9	90.9	1.23×10^9	86.7	4.68×10^8
20	88.7	7.41×10^8	92.8	1.91×10^9	90.4	1.10×10^9	86.5	4.47×10^8
21	88.6	7.24×10^8	92.0	1.58×10^9	90.2	1.05×10^9	86.3	4.27×10^8
22	88.4	6.92×10^8	91.9	1.55×10^9	89.6	9.12×10^8	86.3	4.27×10^8
23	87.7	5.89×10^8	91.8	1.51×10^9	89.4	8.71×10^8	86.1	4.07×10^8
24	87.5	5.62×10^8	91.8	1.51×10^9	89.4	8.71×10^8	85.7	3.72×10^8
25	87.4	5.50×10^8	91.5	1.41×10^9	88.7	7.41×10^8	85.4	3.47×10^8

(continued)

Table 3.4 (continued)

№	Measurement point 1		Measurement point 2		Measurement point 3		Measurement point 4	
	L_{AE}	$10^{(0.1 \cdot L_{AE})}$	L_{AE}	$10^{(0.1 \cdot L_{AE})}$	L_{AE}	$10^{(0.1 \cdot L_{AE})}$	L_{AE}	$10^{(0.1 \cdot L_{AE})}$
26	87.2	5.25×10^8	91.5	1.41×10^9	88.2	6.61×10^8	85.3	3.39×10^8
27	87.2	5.25×10^8	91.5	1.41×10^9	88.1	6.46×10^8	82.7	1.86×10^8
28	87.2	5.25×10^8	91.3	1.35×10^9	87.2	5.25×10^8	81.8	1.51×10^8
29	87.1	5.13×10^8	91.1	1.29×10^9	86.6	4.57×10^8	81.6	1.45×10^8
30	86.5	4.47×10^8	91.0	1.26×10^9	86.4	4.37×10^8	81.5	1.41×10^8
31	86.4	4.37×10^8	90.8	1.20×10^9	85.8	3.80×10^8	81.4	1.38×10^8
32	86.2	4.17×10^8	90.6	1.15×10^9	85.1	3.24×10^8	81.3	1.35×10^8
33	85.9	3.89×10^8	90.5	1.12×10^9	84.9	3.09×10^8	80.6	1.15×10^8
34	85.8	3.80×10^8	90.2	1.05×10^9	84.7	2.95×10^8	80.4	1.10×10^8
35	85.8	3.80×10^8	89.8	9.55×10^8	84.6	2.88×10^8	79.5	8.91×10^7
36	85.8	3.80×10^8	89.7	9.33×10^8	83.8	2.40×10^8	79.2	8.32×10^7
37	85.6	3.63×10^8	89.6	9.12×10^8	82.3	1.70×10^8	78.5	7.08×10^7
38	84.7	2.95×10^8	89.0	7.94×10^8	82.2	1.66×10^8	77.9	6.17×10^7
39	84.6	2.88×10^8	88.7	7.41×10^8	79.3	8.51×10^7	76.1	4.07×10^7
40	84.5	2.82×10^8	88.4	6.92×10^8	79.3	8.51×10^7	75.5	3.55×10^7
41	84.0	2.51×10^8	88.3	6.76×10^8	–	–	73.8	2.40×10^7
42	83.4	2.19×10^8	88.1	6.46×10^8	–	–	73.7	2.34×10^7
43	83.3	2.14×10^8	88.0	6.31×10^8	–	–	66.8	4.79×10^6
44	82.9	1.95×10^8	87.9	6.17×10^8	–	–	–	–
45	82.8	1.91×10^8	87.8	6.03×10^8	–	–	–	–
46	82.8	1.91×10^8	87.3	5.37×10^8	–	–	–	–
47	82.3	1.70×10^8	86.9	4.90×10^8	–	–	–	–
48	82.2	1.66×10^8	86.6	4.57×10^8	–	–	–	–
49	82.1	1.62×10^8	84.8	3.02×10^8	–	–	–	–
50	82.0	1.58×10^8	78.5	7.08×10^7	–	–	–	–
51	81.9	1.55×10^8	75.2	3.31×10^7	–	–	–	–
52	81.8	1.51×10^8	–	–	–	–	–	–
53	81.8	1.51×10^8	–	–	–	–	–	–
54	81.8	1.51×10^8	–	–	–	–	–	–
55	81.7	1.48×10^8	–	–	–	–	–	–
56	81.6	1.45×10^8	–	–	–	–	–	–
57	81.5	1.41×10^8	–	–	–	–	–	–
58	81.5	1.41×10^8	–	–	–	–	–	–
59	81.4	1.38×10^8	–	–	–	–	–	–
60	81.3	1.35×10^8	–	–	–	–	–	–
61	81.3	1.35×10^8	–	–	–	–	–	–
62	81.2	1.32×10^8	–	–	–	–	–	–
63	80.6	1.15×10^8	–	–	–	–	–	–

(continued)

Table 3.4 (continued)

№	Measurement point 1		Measurement point 2		Measurement point 3		Measurement point 4	
	L_{AE}	$10^{(0.1 \cdot L_{AE})}$	L_{AE}	$10^{(0.1 \cdot L_{AE})}$	L_{AE}	$10^{(0.1 \cdot L_{AE})}$	L_{AE}	$10^{(0.1 \cdot L_{AE})}$
64	80.3	1.07×10^8	–	–	–	–	–	–
65	80.2	1.05×10^8	–	–	–	–	–	–
66	80.1	1.02×10^8	–	–	–	–	–	–
67	80.0	1.00×10^8	–	–	–	–	–	–
68	79.9	9.77×10^7	–	–	–	–	–	–
69	79.3	8.51×10^7	–	–	–	–	–	–
70	79.0	7.94×10^7	–	–	–	–	–	–
AM	85.9	7.30×10^8	91.2	1.98×10^9	89.2	1.20×10^9	84.1	4.21×10^8
σ	4.6	–	4.3	–	4.2	–	5.4	–
±Δ	1.1	–	1.2	–	1.3	–	1.6	–
Min	79.0	–	75.2	–	79.3	–	66.8	–
Max	96.4	–	100.9	–	94.8	–	90.6	–

Table 3.5 Sound levels L_{AE} [dBA] at measurement points for nighttime

№	Measurement point 1		Measurement point 2		Measurement point 3		Measurement point 4	
	L_{AE}	$10^{(0.1 \cdot L_{AE})}$	L_{AE}	$10^{(0.1 \cdot L_{AE})}$	L_{AE}	$10^{(0.1 \cdot L_{AE})}$	L_{AE}	$10^{(0.1 \cdot L_{AE})}$
1	90.3	1.07×10^9	87.5	9.04×10^9	89.6	9.12×10^8	78.5	7.08×10^7
2	95.2	3.31×10^9	91.3	1.05×10^9	90.5	1.12×10^9	85.4	3.47×10^8
3	88.6	7.24×10^8	92.4	2.15×10^8	93.6	2.29×10^9	85.5	3.55×10^8
4	95.7	3.72×10^9	89.1	6.1×10^9	88.0	6.31×10^8	87.7	5.89×10^8
5	96.2	4.17×10^9	88.7	1.06×10^9	91.2	1.32×10^9	88.1	6.46×10^8
6	94.8	3.02×10^9	86.4	8.8×10^9	89.6	9.12×10^8	85.0	3.16×10^8
7	96.7	4.68×10^9	90.3	1.62×10^9	92.3	1.70×10^9	86.1	4.07×10^8
8	95.9	3.89×10^9	92.5	2.41×10^9	94.0	2.51×10^9	74.5	2.82×10^7
9	96.5	4.47×10^9	89.7	1.25×10^9	91.6	1.45×10^9	87.2	5.25×10^8
10	101.0	1.26×10^{10}	91.4	1.65×10^{10}	92.3	1.70×10^9	90.6	1.15×10^9
11	–	–	–	–	83.1	2.04×10^8	91.2	1.32×10^9
12	–	–	–	–	92.8	1.91×10^9	88.5	7.08×10^8
13	–	–	–	–	85.1	3.24×10^8	78.0	6.31×10^7
14	–	–	–	–	85.2	3.31×10^8	87.3	5.37×10^8
15	–	–	–	–	85.2	3.31×10^8	86.5	4.47×10^8
16	–	–	–	–	92.7	1.86×10^9	79.8	9.55×10^7
17	–	–	–	–	92.7	1.86×10^9	87.4	5.50×10^8
18	–	–	–	–	102.1	1.62×10^{10}	97.1	5.13×10^9
19	–	–	–	–	94.2	2.63×10^9	88.4	6.92×10^8
20	–	–	–	–	94.2	2.63×10^9	87.0	5.01×10^8
21	–	–	–	–	94.4	2.75×10^9	88.3	6.76×10^8

(continued)

Table 3.5 (continued)

№	Measurement point 1		Measurement point 2		Measurement point 3		Measurement point 4	
	L_{AE}	$10^{(0.1 \cdot L_{AE})}$	L_{AE}	$10^{(0.1 \cdot L_{AE})}$	L_{AE}	$10^{(0.1 \cdot L_{AE})}$	L_{AE}	$10^{(0.1 \cdot L_{AE})}$
22	–	–	–	–	93.2	2.09×10^9	79.2	8.32×10^7
23	–	–	–	–	92.4	1.74×10^9	–	–
24	–	–	–	–	92.7	1.86×10^9	–	–
25	–	–	–	–	88.3	6.76×10^8	–	–
AM	95.1	4.16×10^9	90.2	2.01×10^9	91.2	2.08×10^9	85.8	6.92×10^8
σ	3.3	–	3.5	–	3.9	–	5.0	–
±Δ	2.0		1.5		1.5		2.1	
Min	88.6	–	86.4	–	83.1	–	74.5	–
Max	101.0	–	91.4	–	102.1	–	97.1	–

For the current version of the estimation of the traffic intensity of aircraft, the contour areas of the equivalent sound levels for day and night time were calculated (Table 3.6).

The results in Table 3.6 show that noise levels LA_{eq} and contours with standard values for daytime LA_{eq} = 60 dBA and for night time LA_{eq} = 55 dBA are decisive for the current traffic intensity of aircraft.

Table 3.7 shows the results of calculating the SLAA hazard index from noise pollution by stations (NMPs). It can be seen from them that the levels of noise pollution are within the range 85–170, which corresponds to a moderate level of pollution.

Full-scale acoustic measurements were carried out in the vicinity of the Kyiv International Airport. For the current traffic intensity, contour areas of equivalent sound levels were calculated for day and night. The calculations of noise levels were performed using the Integrated Noise Model (INM) computer program (version 7.0 s), which is recommended by the International Civil Aviation Organization (Figs. 3.21 and 3.22).

Table 3.6 Contour areas of the standard equivalent sound levels for the existing traffic intensity of aircraft

LA_{eq} (dBA)	S (km^2)	
	Day	Night
50	Not standardized	20.17
55	Not standardized	9.02
60	5.35	3.68
65	1.94	Not standardized

Table 3.7 SLAA hazard indices from noise pollution in the vicinity of the Kyiv International Airport

Pollution level indicators	NMP № 1	NMP № 2	NMP № 3	NMP № 4
LA_{eq} (dBA) (day)	59.48	62.43	59.20	54.98
I_{haa}	125	138	124	106
LA_{eq} (dBA) (night)	61.60	61.56	62.56	57.23
I_{haa}	134	134	138	116

3.4 Calculation of the Environmental Risk of the SLAA in the Kyiv City

Kyiv is divided into ten administrative urban districts, formed according to the radial principle and named according to the geographical principle. In accordance with this, the PMP are placed as follows (Table 3.8). According to the CGO, in the Darnytskyi district of Kyiv city, there are no stationary PMPs, therefore, the study took data from the nearest PMP to the district, namely PMP № 3, PMP № 9.

According to statistical data [11], Table 3.9 shows the population of the Kyiv city.

Tables 3.10, 3.11, 3.12 and 3.13 show the calculations of the levels of individual risk of developing undesirable health conditions of the population for the Kyiv city in 2010–2013 period.

The analysis of the SLAA hazard index showed that the risk function is non-linear transformations, it allows to smooth out risks in case of weak impacts, it is a kind of nonlinear filter for identifying dangerous events (Fig. 3.23).

Analyzing the zones of distribution of pollutants in the natural environments of the city of Kyiv, urban planning factors play an important role in their size and configuration [12]. So, in the central part of the Kyiv city, there are weakly functioning "barriers" for the spread of pollutants—green areas. The spatial arrangement of ecosystems of the green zone and hydroecosystems in relation to artificial ones has a combined or multicore and linear character.

Another urban planning factor that adversely affects the spread of pollutants in the natural environment of Kyiv city is the location and quality of the city's transport networks. The high congestion of the main highways of the Kyiv city is due to the lack of an alternative route. Thus, there is no possibility of driving through the residential Darnytsia district to the central part of the city bypassing Demiivska Square, which on weekdays leads to the formation of traffic jams on this road. The emergence of alternative travel options would significantly increase the average vehicle speed [9, 10].

The poor quality of the road surface is observed in almost all areas of the city. Significantly reduces the speed of vehicles, leads to a decrease in road traffic capacity and the formation of traffic congestion and "traffic jams".

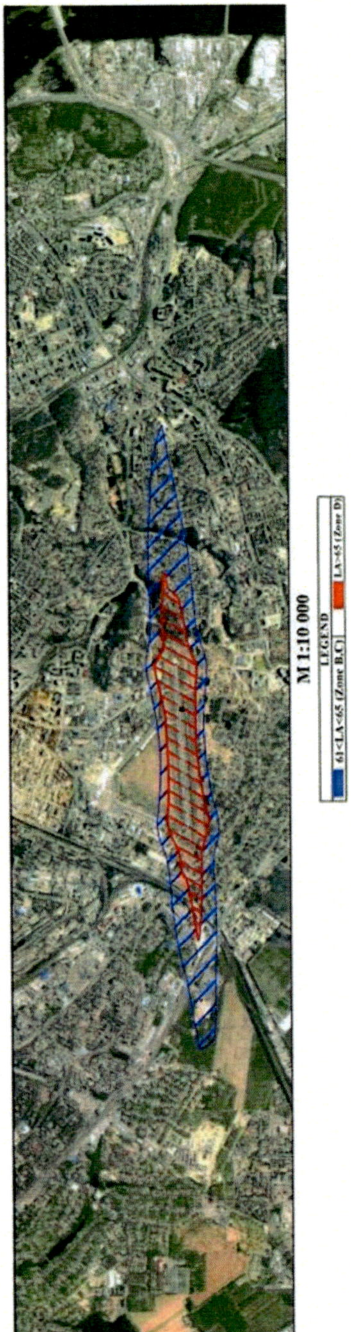

Fig. 3.21 Contours of equivalent sound levels at the current traffic intensity of aircraft of the Kyiv International Airport in the daytime

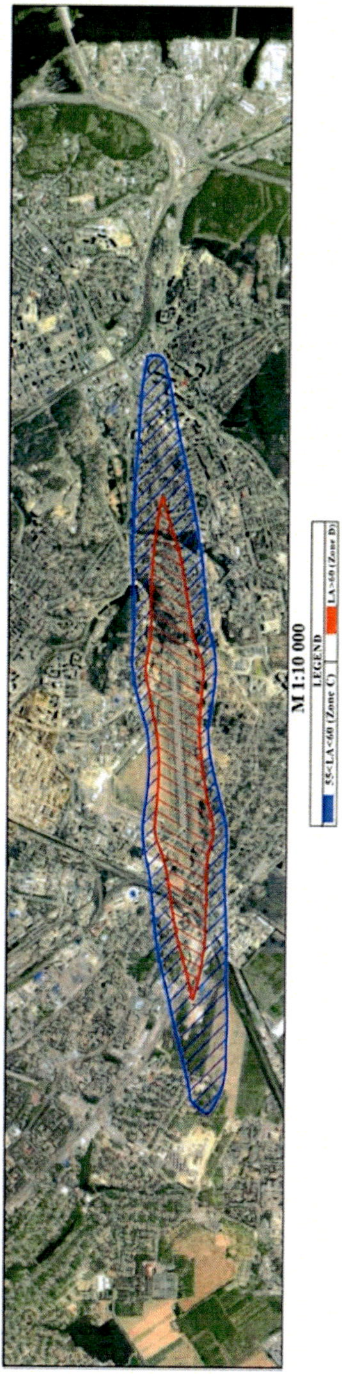

Fig. 3.22 Contours of equivalent sound levels at the current traffic intensity of aircraft of the Kyiv International Airport in the nighttime

Table 3.8 Placement of PMP and NMP in the administrative districts of the Kyiv city

District	PMP and NMP
Holosiiv	PMP № 5, PMP № 13, PMP № 20, NMP № 4
Darnytsia	–
Desna	PMP № 3, PMP № 4
Dnipro	PMP № 9, PMP № 15
Pechersk	PMP № 8
Podil	PMP № 10
Obolon	PMP № 17, PMP № 21
Svyatoshyn	PMP № 11, NMP № 1
Solom'yansk	PMP № 1, NMP № 2, NMP № 3
Shevchenko	PMP № 2, PMP № 6, PMP № 7

Table 3.9 Population of districts of the Kyiv city for 2010–2013

District	Population			
	2010	2011	2012	2013
Holosiiv	235,115	223,064	210,654	205,302
Obolon	315,531	313,612	309,454	306,213
Pechersk	140,454	143,412	146,324	149,514
Podil	191,332	190,991	191,234	190,541
Svyatoshyn	335,898	335,654	335,452	335,226
Solom'yansk	344,716	349,546	353,455	364,371
Shevchenko	230,236	230,567	232,100	231,302
Darnytsia	314,749	314,985	315,289	316,234
Desna	358,353	358,954	366,897	367,118
Dnipro	346,565	348,564	351,325	354,416
Total	2,812,949	2,809,349	2,812,184	2,820,237

Thus, the poor quality of the road surface and the poorly organized function of repairing near-highway roads in the Kyiv city significantly affect the flow of pollutants into the SLAA.

The main objective of measures aimed at reducing environmental risk is to minimize pollution of the SLAA (as the main natural environment that forms the zone of environmental risk) [13, 14].

To achieve this goal on the territory of the Kyiv city, it is necessary to implement 3 main tasks:

- to modern the city's transport networks with an increase in their throughput, quality of road surface, average speed of vehicles;
- to minimize atmospheric emissions from heat and power plants and a number of other industrial facilities;
- to carry out activities for greening the urban space.

Table 3.10 Levels of individual risk of developing an undesirable state of health of the population in 2010

№	Address	Risk level			
		86–170	171–255	256–425	>425
PMP 1	6A, Strazhesko Str.	1.21×10^{-4}	8.30×10^{-5}	5.60×10^{-5}	2.53×10^{-5}
PMP 2	8, Dovzhenko Str.	2.48×10^{-4}	1.73×10^{-4}	1.19×10^{-4}	5.76×10^{-5}
PMP 3	50, Popudrenko Str.	1.40×10^{-4}	9.46×10^{-5}	6.58×10^{-5}	3.09×10^{-5}
PMP 4	2, Lazo Str.	6.95×10^{-5}	4.89×10^{-5}	3.42×10^{-5}	1.76×10^{-5}
PMP 5	37, Nauki Ave.	1.82×10^{-4}	1.17×10^{-4}	7.40×10^{-5}	3.13×10^{-5}
PMP 6	Victory Square	2.42×10^{-4}	1.68×10^{-4}	1.19×10^{-4}	5.96×10^{-5}
PMP 7	Bessarabian Square	4.0×10^{-4}	2.89×10^{-4}	2.11×10^{-4}	1.11×10^{-4}
PMP 8	29, Lesia Ukrainka Boulevard	4.93×10^{-4}	3.41×10^{-4}	2.32×10^{-4}	1.09×10^{-4}
PMP 9	10A, Kaunaska Str.	1.39×10^{-4}	9.38×10^{-5}	6.32×10^{-5}	3.06×10^{-5}
PMP 10	56/60, Mezhyhirska Str.	4.40×10^{-6}	3.70×10^{-6}	2.80×10^{-6}	1.30×10^{-7}
PMP 11	98/2, Victory Avenue	2.34×10^{-4}	1.63×10^{-4}	1.12×10^{-4}	5.42×10^{-5}
PMP 13	Expocenter of Ukraine	1.60×10^{-4}	9.6×10^{-5}	5.87×10^{-5}	2.13×10^{-5}
PMP 15	Hydropark	9.25×10^{-5}	5.26×10^{-5}	3.03×10^{-5}	1.12×10^{-5}
PMP 17	14, Obolonsky Ave	1.63×10^{-4}	1.13×10^{-4}	7.68×10^{-5}	3.60×10^{-5}
PMP 20	Demiivska Square	1.97×10^{-5}	7.59×10^{-6}	3.04×10^{-6}	5.06×10^{-7}
PMP 21	25, Sklyarenka Str.	1.21×10^{-4}	7.81×10^{-5}	5.02×10^{-5}	2.23×10^{-5}

Table 3.11 Levels of individual risk of developing an undesirable state of health of the population in 2011

№	Address	Risk level			
		86–170	171–255	256–425	>425
PMP 1	6A, Strazhesko Str.	1.30×10^{-4}	$8,74 \times 10^{-5}$	6.02×10^{-5}	2.72×10^{-5}
PMP 2	8, Dovzhenko Str.	3.15×10^{-4}	2.20×10^{-4}	1.57×10^{-4}	7.64×10^{-5}
PMP 3	50, Popudrenko Str.	1.38×10^{-4}	8.43×10^{-5}	5.32×10^{-5}	2.00×10^{-5}
PMP 4	2, Lazo Str.	6.94×10^{-5}	4.69×10^{-5}	3.16×10^{-5}	1.43×10^{-5}
PMP 5	37, Nauki Ave.	1.69×10^{-4}	1.13×10^{-4}	7.70×10^{-5}	3.34×10^{-5}
PMP 6	Victory Square	2.82×10^{-4}	1.92×10^{-4}	1.35×10^{-4}	6.55×10^{-5}
PMP 7	Bessarabian Square	4.70×10^{-4}	3.38×10^{-4}	2.45×10^{-4}	1.26×10^{-4}
PMP 8	29, Lesia Ukrainka Boulevard	4.80×10^{-4}	3.22×10^{-4}	2.15×10^{-4}	1.00×10^{-4}
PMP 9	10A, Kaunaska Str.	1.77×10^{-4}	1.19×10^{-4}	8.20×10^{-5}	3.70×10^{-5}
PMP 10	56/60, Mezhyhirska Str.	3.40×10^{-6}	2.70×10^{-6}	1.80×10^{-6}	1.10×10^{-7}
PMP 11	98/2, Victory Avenue	2.53×10^{-4}	1.64×10^{-4}	1.05×10^{-4}	4.68×10^{-5}
PMP 13	Expocenter of Ukraine	1.44×10^{-4}	8.11×10^{-5}	4.45×10^{-5}	1.31×10^{-5}
PMP 15	Hydropark	9.09×10^{-5}	4.96×10^{-5}	2.81×10^{-5}	8.26×10^{-6}
PMP 17	14, Obolonsky Ave	1.96×10^{-4}	1.32×10^{-4}	9.08×10^{-5}	4.10×10^{-5}
PMP 20	Demiivska Square	9.63×10^{-6}	3.70×10^{-6}	1.48×10^{-6}	2.47×10^{-7}
PMP 21	25, Sklyarenka Str.	1.83×10^{-4}	1.27×10^{-4}	9.02×10^{-5}	4.25×10^{-5}

At the same time, it is strategically important to adjust the existing building balance in order to disperse a continuous and rather extensive zone of high technogenic pressure and a higher environmental risk in the city, introducing elements of the ecological framework into the existing building (landscaping, recreational microzones) [15].

3.5 Conclusions

1. The method of selection of the most effective methods of complex reduction of aircraft noise during aircraft operation has been scientifically substantiated. The optimal distribution of aircraft types by routes has been obtained. The distribution of routes was determined in accordance with the data that were in the flight plans. Noise level calculations were performed using the Integrated Noise Model (INM) computer program (version 7.0 s), which is recommended by ICAO.
2. Research results have shown that chemical and physical pollution, which is created by industrial and transport facilities, is an integral part of the total burden on the population. In some cases, it exceeds hygienic standards and, as a result, can negatively affect the state of human health. To prevent the harmful effects of chemical and physical pollution on the population, it is necessary to create a system of protective measures to reduce levels in the territories of populated areas adjacent to the airport, including the establishment of sanitary protection zones and zones of restriction of development in places where the population is located.
3. To determine the boundaries of sanitary protection zones and zones of restriction of development in the locations of existing and projected facilities of the Kyiv International Airport, it is necessary to use theoretical (calculated) methods for determining the hazard index of levels of chemical and physical pollution, which with sufficient accuracy allows to determine the hazard levels at any distance and height from the object. The results of the research showed that the levels that were determined by theoretical research methods coincide with those that were determined by the measuring device. This indicates that both methods are valid and can be used in a hygienic assessment of the situation. But the instrumental method does not make it possible to establish zones of development restriction, and therefore it is used as a method that verifies theoretical calculations.
4. The zones of distribution of pollutants in the natural environments of the city of Kiev are analyzed. Urban planning factors play an important role in their size and configuration.

Table 3.12 Levels of individual risk of developing an undesirable state of health of the population in 2012

№	Address	Risk level			
		86–170	171–255	256–425	>425
PMP 1	6A, Strazhesko Str.	1.88×10^{-4}	1.31×10^{-4}	9.25×10^{-5}	4.35×10^{-5}
PMP 2	8, Dovzhenko Str.	3.34×10^{-4}	2.35×10^{-4}	1.69×10^{-4}	8.46×10^{-5}
PMP 3	50, Popudrenko Str.	1.36×10^{-4}	9.61×10^{-5}	6.97×10^{-5}	3.58×10^{-5}
PMP 4	2, Lazo Str.	6.72×10^{-5}	4.71×10^{-5}	3.36×10^{-5}	1.63×10^{-5}
PMP 5	37, Nauki Ave.	2.00×10^{-4}	1.33×10^{-4}	8.78×10^{-5}	3.94×10^{-5}
PMP 6	Victory Square	3.50×10^{-4}	2.54×10^{-4}	1.87×10^{-4}	1.01×10^{-4}
PMP 7	Bessarabian Square	5.42×10^{-4}	4.15×10^{-4}	3.17×10^{-4}	1.90×10^{-4}
PMP 8	29, Lesia Ukrainka Boulevard	5.49×10^{-4}	3.80×10^{-4}	2.58×10^{-4}	1.21×10^{-4}
PMP 9	10A, Kaunaska Str.	2.23×10^{-4}	1.60×10^{-4}	1.13×10^{-4}	5.98×10^{-5}
PMP 10	56/60, Mezhyhirska Str.	3.10×10^{-6}	2.10×10^{-6}	1.10×10^{-6}	2.10×10^{-7}
PMP 11	98/2, Victory Avenue	2.78×10^{-4}	2.06×10^{-4}	1.53×10^{-4}	8.39×10^{-5}
PMP 13	Expocenter of Ukraine	1.03×10^{-4}	5.95×10^{-5}	3.42×10^{-5}	1.08×10^{-5}
PMP 15	Hydropark	8.11×10^{-5}	4.61×10^{-5}	2.66×10^{-5}	9.79×10^{-6}
PMP 17	14, Obolonsky Ave	2.58×10^{-4}	1.85×10^{-4}	1.31×10^{-4}	6.91×10^{-5}
PMP 20	Demiivska Square	1.26×10^{-5}	5.06×10^{-6}	2.21×10^{-6}	3.16×10^{-7}
PMP 21	25, Sklyarenka Str.	2.16×10^{-4}	1.60×10^{-4}	1.15×10^{-4}	6.21×10^{-5}

Table 3.13 Levels of individual risk of developing an undesirable state of health of the population in 2013

№	Address	Risk level			
		86–170	171–255	256–425	>425
PMP 1	6A, Strazhesko Str.	1.45×10^{-4}	1.01×10^{-4}	6.94×10^{-5}	3.37×10^{-5}
PMP 2	8, Dovzhenko Str.	3.15×10^{-4}	2.21×10^{-4}	1.53×10^{-4}	7.66×10^{-5}
PMP 3	50, Popudrenko Str.	1.71×10^{-4}	1.23×10^{-4}	8.67×10^{-5}	4.34×10^{-5}
PMP 4	2, Lazo Str.	6.89×10^{-5}	4.66×10^{-5}	3.14×10^{-5}	1.52×10^{-5}
PMP 5	37, Nauki Ave.	3.10×10^{-4}	2.70×10^{-4}	1.60×10^{-4}	3.10×10^{-5}
PMP 6	Victory Square	3.28×10^{-4}	2.37×10^{-4}	1.73×10^{-4}	9.12×10^{-5}
PMP 7	Bessarabian Square	4.55×10^{-4}	3.38×10^{-4}	2.52×10^{-4}	1.35×10^{-4}
PMP 8	29, Lesia Ukrainka Boulevard	5.05×10^{-4}	3.49×10^{-4}	2.38×10^{-4}	1.11×10^{-4}
PMP 9	10A, Kaunaska Str.	1.55×10^{-4}	1.11×10^{-4}	8.07×10^{-5}	4.14×10^{-5}
PMP 10	56/60, Mezhyhirska Str.	3.10×10^{-6}	2.10×10^{-6}	1.10×10^{-6}	2.10×10^{-7}
PMP 11	98/2, Victory Avenue	2.39×10^{-4}	1.72×10^{-4}	1.25×10^{-4}	6.40×10^{-5}
PMP 13	Expocenter of Ukraine	9.25×10^{-5}	4.80×10^{-5}	2.49×10^{-5}	7.11×10^{-6}

(continued)

Table 3.13 (continued)

№	Address	Risk level			
		86–170	171–255	256–425	>425
PMP 15	Hydropark	7.28×10^{-5}	3.91×10^{-5}	2.16×10^{-5}	6.74×10^{-6}
PMP 17	14, Obolonsky Ave	2.27×10^{-4}	1.59×10^{-4}	1.10×10^{-4}	5.52×10^{-5}
PMP 20	Demiivska Square	2.88×10^{-4}	2.07×10^{-4}	1.46×10^{-4}	7.31×10^{-5}
PMP 21	25, Sklyarenka Str.	1.30×10^{-4}	9.09×10^{-5}	6.49×10^{-5}	3.15×10^{-5}
NMP 1	–	2.60×10^{-5}	2.20×10^{-5}	1.20×10^{-5}	2.60×10^{-6}
NMP 2	–	3.10×10^{-5}	2.10×10^{-5}	1.00×10^{-5}	3.10×10^{-6}
NMP 3	–	5.20×10^{-5}	4.20×10^{-5}	2.20×10^{-5}	5.20×10^{-6}
NMP 4	–	4.40×10^{-5}	2.40×10^{-5}	1.40×10^{-5}	4.40×10^{-6}

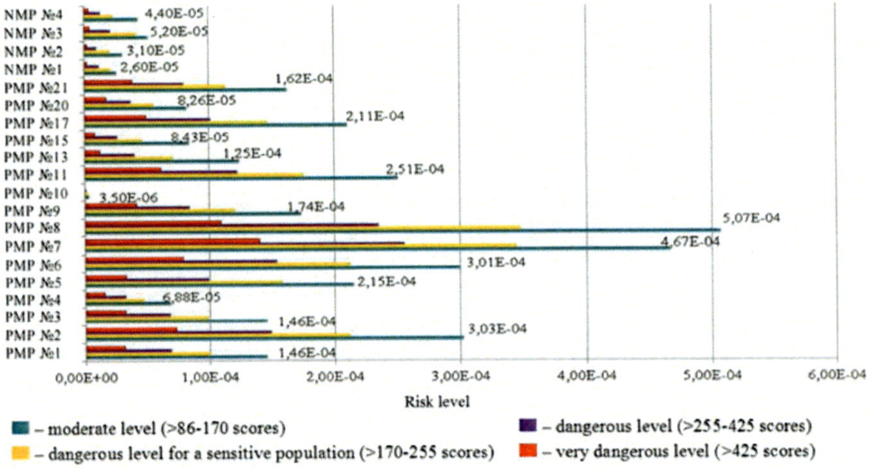

Fig. 3.23 Levels of individual risk of developing an undesirable state of health of the population at the NMPs and PMP2 of Kyiv city

References

1. Bazylevych, V., Kupalova, G.: Climate change in Kyiv: ways to conteract and minimize negative effects. Bulletin of Taras Shevchenko National University of Kyiv. Economics **11** (152), 6–12
2. Danylo, O., Bechtel, B., Schepaschenko, D., Fritz, S.: Contributing to WUDAPT: a local climate zone classification of two cities in Ukraine. IEEE J. Sel. Topics Appl. Earth Obs. Remote Sens. **9**(5), 1841–1853 (2016). https://doi.org/10.1109/JSTARS.2016.2539977
3. GOST 20 444-85: Noise, Transport streams in settlements, Method for determining the noise characteristics. Stroyizdat, Moscow (1986). (in Russian)
4. SSR № 173: State Sanitary Rules of Planning and Development of Settlements. Ministry of Construction and Architecture, Kyiv (2003)
5. Isaenko, V., Zaporozhets, A., Babikova, K., Gulevets, D., Savchenko, S.: Review of methods and means on monitoring the air pollution. Proc. Natl. Aviat. Univ. **3**(80), 61–70 (2019). https://doi.org/10.18372/2306-1472.80.14275

6. Zaporozhets, A., Babak, V., Isaienko, V., Babikova, K.: Analysis of the air pollution monitoring system in Ukraine. In: Babak V., Isaienko V., Zaporozhets, A. (eds.) Systems, Decision and Control in Energy I. Studies in Systems, Decision and Control, vol. 298, pp. 85–110. Springer, Cham (2020). https://doi.org/10.1007/978-3-030-48583-2_6
7. Levchenko, L.A., Sidko, O.S.: GIS-technology application for the visualization the modeling results of the spreading aircraft noise. Control, Navig. Commun. Syst. **2**(48), 120–125 (2018). https://doi.org/10.26906/SUNZ.2018.2.120
8. Zhdanko, I.M., Zinkin, V.N., Soldatov, S.K., Bogomolov, A.V., Sheshegov, P.M.: Fundamental and applied aspects of preventing the adverse effects of aviation noise. Human Physiol. **42**, 705–714 (2016). https://doi.org/10.1134/S0362119716070227
9. Movchan, Y.I., Rybalova, O.V., Gulevets, D.V.: Environmental risk of deterioration of the modern condition of urban areas. Eastern-European J. Enterpr. Technol. **11**(63), 37–41 (2013). https://doi.org/10.15587/1729-4061.2013.14594
10. Gulevets, D.V.: The tools for improving comfort urban areas by a comprehensive indicator of environmental hazards. Probl. Environ. Biotechnol. **4**(2) (2013). https://doi.org/10.18372/2306-6407.2.5505
11. Lesnik, O.M., Girs, O.A.: The analysis of providing of Kyiv population by green plantations. Ukrainian J. For. Wood Sci. **216**(1), 15–21 (2015)
12. Popov, O.O., Iatsyshyn, A.V., Kovach, V.O., Artemchuk, V.O., Kameneva, I.P., Taraduda, D.V., Sobyna, V.O., Sokolov, D.L., Dement, M.O., Yatsyshyn, T.M.: Risk assessment for the population of Kyiv, Ukraine as a result of atmospheric air pollution. J. Health Pollut. **10**(25), 200303 (2020). https://doi.org/10.5696/2156-9614-10.25.200303
13. Zou, B., Li, S., Lin, Y., Wang, B., Cao, S., Zhao, X., Peng, F., Qin, N., Guo, Q., Feng, H., Matthew, C.J., Xu, S., Duan, X.: Efforts in reducing air pollution exposure risk in China: State versus individuals. Environ. Int. **137**, 105504 (2020). https://doi.org/10.1016/j.envint.2020.105504
14. Fann, N., Roman, H.A., Fulcher, C.M., Gentile, M.A., Hubbell, B.J., Wesson, K., Levy, J.I.: Maximizing health benefits and minimizing inequality: incorporating local-scale data in the design and evaluation of air quality policies. Risk Anal. **31**(6), 908–922 (2011). https://doi.org/10.1111/j.1539-6924.2011.01629.x
15. Polzonetti, A., Sagratella, M.: Smart city and green development. In: Al-Sharhan, S., et al. (eds.) Challenges and Opportunities in the Digital Era. I3E 2018. Lecture Notes in Computer Science, vol. 11195, pp. 191–204. Springer, Cham (2018). https://doi.org/10.1007/978-3-030-02131-3_18

Chapter 4
Ecological and Energy Analysis of the Green Areas and the Surface Layer of Atmospheric Air in the Districts of the Kyiv City

Abstract In this chapter, the calculation of the variant of the emergetic balance between green areas (natural component) and technogenic components (cost component) of the Kyiv city's districts is carried out. The urboecosystem and its ordering can be considered as a result of plant photosynthesis and the life-sustaining ability of environmental factors together with the consumption of non-renewable resources and the further production of goods and services. Green area energy indicators were calculated for city districts in order to assess the balance of energy consumption and production within a year. Emergy is one kind of available energy that has been previously used directly or indirectly to produce a product or service. The following were used as indicators for the emergetic balance of the expenditure component: consumed energy, calculated in terms of the equivalent fuel, and energy of noise and electromagnetic radiation.

Keywords Urban ecosystem · Emergy · Green areas · Energy of noise and electromagnetic radiation

4.1 Analysis and Assessment of the State of the Green Area of the Kyiv City

An analysis of the process of urban development of landscapes showed that 4 landscape and architectural systems of ancient and modern buildings, parks and forest parks have developed on the Kyiv territory.

The largest share of the green zone is coniferous and mixed forests—327 km^2 (39%), found in the eastern, north-western and western parts of the city and are mainly of artificial origin. Deciduous forests cover an area of 42 km^2 (5%). Of these, riverine poplar and willow forests on sandy terraces and the Dnieper islands occupy 15 km^2 (2%), hornbeam-oak forests of the forest-steppe part in the south-west of the city—20 km^2 (2%), other types of deciduous forests (ash-oak, floodplain oak forests, alder)—7 km^2 or 1% of the area [1, 2]. Floodplain shrub groups and meadows make up 58 km^2 (7%). The area of hydrotopes is 41 km^2 (5%—the area of the

D. Gulevets et al., *Environmental Control for Ensuring Cities Safety*,
Lecture Notes in Intelligent Transportation and Infrastructure,
https://doi.org/10.1007/978-3-030-66710-8_4

Dnipro) and 21 km^2 (2%—stagnant water bodies). Agroecosystems, parks and squares cover only 33 km^2 (4%); technoecosystems include residential (171 km^2, 21%), industrial (86 km^2, 10%) zones, transport networks (56 km^2, 7%) [2].

Basically, the increase in the territory of the city occurs due to deforestation. According to the accepted classification, green areas are functionally divided into three main groups: (a) green areas of common use; (b) green areas of limited use—in residential areas, land-use areas of children's institutions, schools, universities, scientific and medical institutions, enterprises and organizations; (c) green areas for special purposes—sanitary protection zones of industrial enterprises, water protection, reclamation and hail protection belts, greenhouses and nurseries, corridors of engineering networks, green areas of streets, roads and cemeteries [3, 4].

4.1.1 Common Areas

Based on the functional-spatial analysis of common areas (calculation of the areas of objects on a topographic basis within the limits established by the reporting materials, determination of the functional profile and categorization of objects and their field survey), a number of disagreements were identified, namely: inconsistency of the available area, the actual absence of individual territories as objects of green areas, belonging to other functional groups and categories (special purpose and limited use) [5].

Common areas include:

- parks (hydroparks, meadow parks, forest parks, parks of culture and recreation, parks monuments of landscape gardening, sports, children's, historical, national, memorial and others), recreational areas, gardens, squares and playgrounds;
- monuments of cultural and historical heritage;
- squares, boulevards, avenues;
- streets, roads, lanes, driveways, pedestrian and bicycle paths;
- beaches;
- cemetery;
- other common areas.

The list of available (geographically and typologically defined) green areas of common use, which should be the basis for the subsequent inventory and planning fixation, is given in Table 4.1.

The study of the dynamics of the contours of green spaces was carried out using the material of the geological portal "Urban Cadastre of Kyiv". Space images of green areas of the city of Kyiv were analyzed for 2013. Based on the results of the analysis, the development of existing green spaces and territories was identified, where it was possible to create new landscaping objects [6, 7]. The results of the analysis and assessment of the network, the state of green areas of public use allow to determine the actual indicators of security in the city (16.5 m^2/person) and administrative districts, and are given in Table 4.2.

Table 4.1 Green areas of common use of the Kyiv city (according to the reporting data of the "Kyivzelenbud" for 2013)

Administrative districts	Parks of culture and rest (ha)	Recreation parks (ha)	Squares (ha)	Boulevards (ha)	Landscaped plantings (ha)	Specialized parks (ha)	Total green areas of common use (ha)	Supply indicators (m^2/person)
Holosiiv	140.9	192.49	15.36	22.79	472.35	485.1	1328.99	64.7
Darnytsia	110.22	22.04	332.48	51.2	–	–	515.94	17.9
Desna	–	278.71	17.16	45.9	2.88	–	344.65	10.2
Dnipro	430.3	590.19	36.1	28.93	–	–	1085.52	32.9
Obolon	–	39.24	128.86	103.71	–	127.5	271.81	8.9
Pechersk	43.75	183.46	12.18	18.91	98.51	130.2	487.01	37.5
Podil	–	150.52	22.51	22.87	68.85	–	264.75	14.6
Svyatoshyn	–	65.76	70.52	59.32	101.66	–	297.26	9.4
Solom'yansk	–	269.24	39.67	24.25	102.8	–	435.96	15
Shevchenko	–	324.23	50.57	8.22	–	62	445.02	19.1
Total	725.17	2115.88	725.41	386.1	847.05	804.8	5604.41	21.4

Table 4.2 Actual provision of available green areas of common use

Administrative districts	Territory of green areas of common use (ha)	Provision of green areas for general use (m^2/person)
Holosiiv	882.15	43
Darnytsia	173.67	6
Desna	351.61	10.4
Dnipro	1177.84	35.6
Obolon	825.74	15.1
Pechersk	360.33	27.7
Podil	216.95	12
Svyatoshyn	211.48	6.7
Solom'yansk	241.76	8.3
Shevchenko	454	19.4
Total	4337.63	16.5

According to [8], the zone of green areas of general use for cities with a population of more than 1 million people should be at least 10–11 m^2/person.

So, the actual provision with green areas of common use in Darnytsia, Svyatoshyn and Solom'yansk districts is lower than the standardized one.

In the Darnytsia district, the greatest discrepancies in the areas of the considered territories and their functional distribution are in squares, boulevards and avenues, especially a large area of 301.3 hectares, which was classified as a public garden—this is the «Bortnichi village» object, which is not a green area of public use. Most green areas require landscape reconstruction.

Inventory areas in the Dnipro district of Kyiv city, with the exception of minor differences, are most consistent with reality in the category of green areas of general use. Geographically, Trukhanov Island (508 hectares) belongs to the Dnipro region, which significantly influenced the increase in the provision of normative green areas. In recent years, the quality of the plantings has improved significantly, but most parks, squares, boulevards require reconstruction.

The Desna district has the most well-maintained inventory of parks and squares. Data on boulevards, avenues and landscaped public plantings have significant discrepancies in areas and distribution by functional groups. This applies to Brovarskyi avenue, Roman Shukhevych avenue, Mayakovsky avenue, etc.

The quality condition of the existing green areas of common use of Obolon district is satisfactory. Some of the plots of limited use (green areas of residential development on Roman Shukhevych St., Heroes of Dnipro St., etc.) and special purposes (areas along Bogatyrskaya St., Obolonsky Av., Roman Shukhevych St., Roman Shukhevych Av.) are erroneously classified as green areas of public use.

In Khreshchaty Park, Mariinsky Park and in the City Garden of the Pechersky District, reconstruction work is almost completed, the state of the remaining green areas is satisfactory. There are differences in the inventory and analytical data of a number of territories and cases of attributing limited-use and special-purpose plantings to squares and boulevards (transport interchanges near the Paton bridge, on Kikvidze street, near the Pechersky bridge, etc.).

Svyatoshyn district has a low level of provision with green areas of common use (6.7 m^2/person). The discrepancy between the available areas of public facilities is almost 100 hectares. In a number of cases, green areas of special purpose are referred to squares, mainly within the limits of transport interchanges. The state of the green areas is satisfactory, but they all require landscape reconstruction.

There are differences in the area of green spaces in the Solom'yansk district. Many protective plantings are classified as public areas (along the railway in the area of Karavaevi dachi, on Liniyna, Lokomotyvna, Kropyvnytska Streets), as well as some green areas of limited use—on Socialistychna St., on Yangel St., a memorial at the Solom'yansk cemetery.

The green areas of public use of the Shevchenko district are well-groomed, although the vast majority require landscape reconstruction, and a positive exception is the T. Shevchenko park and a number of squares. Inventory data are as close as possible to the actual for the main groups of territories [2].

4.1.2 Territories of Limited Use

Plantations of limited use of the Kyiv city occupy an area of 11,638.6 hectares and are an important component of the green areas of the city. The largest and most important share in this type of plantings is green areas in residential areas. They consist of intra-quarter plantings and adjoining strips [9].

Plantations in residential areas are heterogeneous, but there is a certain regularity in their distribution both in terms of aesthetic value, species composition, and the attitude of land users and district administrations.

Areas of earlier development (residential areas Voskresenka, Stara Darnytsa, Otradnyi, Lisovyi, Borshagivka) are characterized by a rich assortment of tree and shrub species, but they lose in aesthetic perception, because the period of settlement of these residential areas was accompanied by active amateur activity of residents, as a result of which randomness was allowed and planting density (as a rule, fruit trees and shrubs were planted). For plantings on the territories of schools and child care centers, typical problems are typical for plantings in residential areas, with the only difference that the plantings of this group withstand a more intense load, and therefore require frequent repair and careful maintenance.

The main purpose of plantings of industrial enterprises is to fulfill their sanitary and recreational function. The percentage of their greening is mainly 15–20% of the total area. The territories of public institutions, trade enterprises and hospitals have always received sufficient attention. Their condition is approximately at the same level as before (Table 4.3) [2].

Table 4.3 Zones of green areas of limited use of the Kyiv city (ha) (according to the reporting data of "Kyivzelenbud")

Administrative districts	Total	Residential areas			Pre-schools and schools	Universities, scientific and other institutions	Hospitals	Enterprises and other institutions
		All	Urban high-rise	Private				
Holosiiv	1573.7	603	157	446	78	484.7	46.6	361.4
Darnytsia	1298.4	631.2	329.9	301.3	73	–	2.2	160
Desna	968.3	416.7	377.6	39.1	81.4	28.4	11.2	430.6
Dnipro	1144.8	507	400.5	106.5	295.7	80.5	26.7	14.9
Obolon	1129.9	464.7	321.7	143	103.2	7.1	18.5	396.7
Pechersk	325	206.7	77.6	129.1	28.2	1.9	0.2	88.9
Podil	638.1	355	181	174	44.3	1.8	44.7	51.8
Svyatoshyn	1384.9	626.1	465	161.1	39.7	–	64	497.8
Solom'yansk	2589.4	1128.5	807.8	320.7	93.2	638.1	59.5	601.5
Shevchenko	585.9	472.5	346	126.5	28.7	–	11	60.9
Total	11,638.6	5411.4	3464.1	1947.3	865.4	1242.5	284.6	2664.5

4.1.3 Territories of Special Use

Special-purpose green spaces—plantings along streets, in sanitary protection and security zones, in the territories of botanical and zoological gardens, exhibitions, cemeteries and crematoria, high voltage power lines; forest reclamation plantings; plantings of nurseries, floriculture farms; roadside plantings within cities and other settlements [10, 11].

Plantings for special purposes are very different in nature and function. Street plantations, consisting mainly of row plantings of trees, and, as well as plantations of common use, play a significant health-improving and climatic-regulatory role. Their total area is 784.6 hectares (Table 4.4).

Requirements for plantings for special purposes:

- restoration of a single «green belt» around the city;
- ensuring a continuous honeycomb effect of various elements of the city's greening system;
- provision of a local climatic factor (draft, shade, high humidity);
- preference for plantings with increased absorption capacity at industrial facilities of the city;
- active introduction of elements of mobile and vertical gardening in areas where there are no open soil areas;
- compliance with the normative indicators of landscaping in order to prevent overloading with their elements (ensuring the unhindered movement of pedestrians, proper visibility of traffic control elements, etc.);
- preference for lawn coverage.

Functional features of special-purpose plantings:

- noise and sawn-off;
- aesthetically-decorative (harmonious combination of the shape and color of the plantings, taking into account the seasonal change in color);

Table 4.4 Actual provision of available green areas for special use

Administrative districts	Territory of green areas for special use (ha)
Darnytsia	64.1
Desna	74.3
Dnipro	53.1
Obolon	99.9
Pechersk	59.0
Podil	50.3
Svyatoshyn	131.4
Solom'yansk	87.4
Shevchenko	93.0
Total	784.6

- absorbing (the ability to absorb pollutants and harmful emissions in the SLAA);
- protective.

4.2 Assessment of the Emergetic Potential of Green Areas in the Kyiv City

On the basis of the methodology developed by Odum [12], the indicators of the green zone's energy were calculated for the districts of the Kyiv city. The basic concepts of emergetic analysis are "emergy" and "transformation index".

Emergia is one kind of available energy that was previously used directly or indirectly to produce a product or service [13–15]. If energy is the potential ability to do work, and it comes from both natural and economic sources that may be located far from the city, then energy is the sum of all energy that was used directly or indirectly to create a resource, product or service., expressed in units of one type of energy, namely the energy of the sun (seJ) [16].

For calculating the emergy of the green area, it is necessary to take into account the energy of sunshine that was spent on photosynthesis, the energy of consumed water and the energy of the soil, and multiply these indicators by the corresponding transformation indices (Table 4.5).

The emergy of green spaces in Holosiiv district is

$$6.48 \times 10^{14} + 1.63 \times 10^{18} + 2.10 \times 10^{16} = 1.60 \times 10^{18}\ \text{seJ}.$$

Noise load. To assess the energy potential of the noise load, the Weber-Fechner law was taken, according to which the sensitivity of the hearing aid is proportional to the logarithm of the ratio of the intensity of sounds I and I_0 (formula 2.8, Chap. 2 of this book).

According to research [17] and own research [2, 18], a table of average values of equivalent annual noise levels for the districts of the Kyiv city in the day and night periods was formed, which are given in Table 4.6.

Electromagnetic radiation. To assess the energy of electromagnetic radiation, data from the dissertation research by Dumansky [19] were used. According to the calculations of the authors of the study, the electromagnetic load did not exceed the established limits and did not constitute a threat to the population (Table 4.7).

The assessment of the energy potential of the electromagnetic load was carried out taking into account the type of source and the number of their placement within the region and the measured level of the electromagnetic load.

Consumed energy. The use of energy materials and oil refined products for industrial and domestic needs is defined as the sum of data from enterprises on the consumption of these resources for conversion into other types of fuel and energy, for non-energy purposes (as raw materials, material), for final consumption, and their losses during distribution, transportation and storage, regardless of the sources

Table 4.5 Energy balance of green areas of Kyiv city's districts

District	Green areas (m^2)	Solar energy	Solar emergy	Water energy	Water emergy	Soil energy	Soil emergy	General emergy
Holosiiv	7.58×10^{7}	2.00×10^{15}	2.00×10^{15}	9.93×10^{13}	1.53×10^{18}	1.03×10^{12}	6.48×10^{16}	1.6×10^{18}
Darnytsia	1.91×10^{8}	5.04×10^{15}	5.04×10^{15}	2.50×10^{14}	3.86×10^{18}	2.59×10^{12}	1.63×10^{17}	4.0×10^{18}
Desna	2.21×10^{7}	5.83×10^{14}	5.83×10^{14}	2.90×10^{13}	4.47×10^{17}	3.00×10^{11}	1.89×10^{16}	4.6×10^{17}
Dnipro	1.81×10^{7}	4.79×10^{14}	4.79×10^{14}	2.38×10^{13}	3.66×10^{17}	2.46×10^{11}	1.55×10^{16}	3.8×10^{17}
Obolon	1.66×10^{7}	4.39×10^{14}	4.39×10^{14}	2.18×10^{13}	3.36×10^{17}	2.25×10^{11}	1.42×10^{16}	3.5×10^{17}
Pechersk	7.62×10^{6}	2.01×10^{14}	2.01×10^{14}	9.98×10^{12}	1.54×10^{17}	1.03×10^{11}	6.51×10^{15}	1.6×10^{17}
Podil	1.85×10^{7}	4.89×10^{14}	4.89×10^{14}	2.43×10^{13}	3.75×10^{17}	2.51×10^{11}	1.58×10^{16}	3.9×10^{17}
Svyatoshyn	1.50×10^{8}	3.96×10^{15}	3.96×10^{15}	1.97×10^{14}	3.03×10^{18}	2.04×10^{12}	1.28×10^{17}	3.1×10^{18}
Solom'yansk	3.13×10^{7}	8.25×10^{14}	8.25×10^{14}	4.10×10^{13}	6.32×10^{17}	4.24×10^{11}	2.67×10^{16}	6.5×10^{17}
Shevchenko	1.18×10^{7}	3.12×10^{14}	3.12×10^{14}	1.55×10^{13}	2.39×10^{17}	1.60×10^{11}	1.01×10^{16}	2.4×10^{17}
Total	5.17×10^{8}	1.43×10^{16}	1.43×10^{16}	7.11×10^{14}	1.10×10^{19}	7.36×10^{12}	4.64×10^{17}	1.1×10^{19}

Table 4.6 Values of equivalent noise levels for Kyiv city's districts

District	LA_{eq} (dBA)	
	Day	Night
Holosiiv	73.2	45.9
Darnytsia	70.1	43.3
Desna	74.2	45.6
Dnipro	69.0	42.4
Obolon	74.5	43.5
Pechersk	70.1	41.1
Podil	66.4	40.3
Svyatoshyn	72.9	44.2
Solom'yansk	71.2	42.1
Shevchenko	70.1	40.1

of income (they were purchased at their own expense, or they came as raw materials supplied by the customer) [20–23].

Conventional fuel is a unit of accounting for fossil fuels, which is used to compare the efficiency of various types of fuel and their total accounting. As a unit of [t.c.f.] 1 kg of fuel with a calorific value of 7000 kcal/kg (29.3 MJ/kg) is taken.

The ratio between [t.c.f.] and natural fuel is expressed by the formula

$$B_y = \frac{Q_H^P}{7000} \times B_H = E \times B_H,$$

where B_y—mass of equivalent amount of equivalent fuel (kg); B_H—mass of natural fuel (kg) (solid and liquid fuel) or (m^3) (gaseous); Q_H^P—net calorific value of a given natural fuel (kcal/kg) or (kcal/m^3); $E = \frac{Q_H^P}{7000}$—calorie equivalent.

The conversion of the amount of fuel of this type into the conditional one is performed using a coefficient equal to the ratio of the heat content of 1 kg of fuel of this type to the heat capacity of 1 kg of the conventional fuel (Table 4.8).

According to the Table 3.8 of the Chap. 3 of this book, the total population in 2013 was 2,820,237 people, so for one person on average is 6,266,083/ 2,820,237 = 2.21 t.c.f./person.

Table 4.9 shows the calculations of the emergetic balance of the green areas of Kyiv and their relation to the expenditure component.

Analyzing the indicators of the Table 4.9, it should be noted that the share of the emergetic potential of the green areas of the city districts is 1.61×10^{17} to 4.02×10^{18} seJ.

Consumption indicators of emergy in general for the districts of the Kyiv city are higher and amount to 3.81×10^{17} to 4.02×10^{18} seJ, which is higher than the indicators of the green component; consumed energy (3.81×10^{17} to 4.02×10^{18} seJ) and noise load (1.08×10^{13} to 1.49×10^{14} seJ) have high indicators.

Table 4.7 Results of determining the levels of electromagnetic pollution and its load on the population of the administrative districts of the Kyiv city

District	EMR source	EMR objects' amount	MPL	Estimated EMR level	Measured EMR level	EMR load	
						From a separate source	General
Shevchenko	KTTs	1	3 W/m	0.6 W/m	0.7 W/m	0.233	0.923
	BS	100	2.5 μW/cm^2	0.2 μW/cm^2	0.5 μW/cm^2	0.377	
	Ratel	1	3 W/m	1.06 W/m	1.5 W/m	0.500	
Pechersk	KTTs	1	3 W/m	0.11 W/m	0.8 W/m	0.287	0.91
	BS	120	2.5 μW/cm^2	0.1 μW/cm^2	0.3 μW/cm^2	0.623	
Podil	KTTs	1	3 W/m	0.6 W/m	0.9 W/m	0.3	0.60
	BS	100	2.5 μW/cm^2	0.1 μW/cm^2	0.04 μW/cm^2	0.299	
Obolon	KTTs	1	3 W/m	0.7 W/m	0.9 W/m	0.30	0.38
	BS	90	2.5 μW/cm^2	0.5 μW/cm^2	0.45 μW/cm^2	0.083	
Svyatoshyn	KTTs	1	3 W/m	0.8 W/m	0.9 W/m	0.30	0.21
	BS	80	2.5 μW/cm^2	0.1 μW/cm^2	0.11 μW/cm^2	0.081	
	LEP-330	1	1 kW/m	0.2 kW/m	0.1 kW/m	0.10	
	Kyiv AVO	1	15 W/m	0.7 W/m	0.5 W/m	0.03	
Solom'yansk	KTTs	1	3 W/m	1.3 W/m	0.93 W/m	0.31	0.95
	BS	125	2.5 μW/cm^2	0.1 μW/cm^2	0.1 μW/cm^2	0.32	
	RP-3G	1	2.5 μW/cm^2	0.6 μW/cm^2	0.2 μW/cm^2	0.08	
	DPRM-261	1	25 W/m	2.7 W/m	0.2 W/m	0.007	
	BPRM-261	1	25 W/m	0.5 W/m	0.1 W/m	0.033	
	KRM-261	1	3 W/m	1.7 W/m	0.1 W/m	0.033	
	GRM-261	1	25 μW/cm^2	0.8 W/m	0.01 W/m	0.003	
	DPSP	1	3 W/m	0.2 W/m	0.2 W/m	0.067	
	PRTs	1	4–25 W/m	1.5 W/m	0.3 W/m	0.10	
Holosiiv	KTTs	1	3 W/m	0.2 W/m	0.01 W/m	0.033	0.22
	BS	70	2.5 μW/cm^2	0.2 μW/cm^2	0.2 μW/cm^2	0.127	
	Meteo RLS	1	10 μW/cm^2	0.2 μW/cm^2	0.1 μW/cm^2	0.01	
	RZ	1	6 W/m	0.2 W/m	0.1 W/m	0.083	
Desna	KTTs	1	3 W/m	0.5 W/m	0.5 W/m	0.107	0.15
	BS	60	2.5 μW/cm^2	0.1 μW/cm^2	0.1 μW/cm^2	0.041	

(continued)

Table 4.7 (continued)

District	EMR source	EMR objects' amount	MPL	Estimated EMR level	Measured EMR level	EMR load	
						From a separate source	General
Dnipro	KTTs	1	3 W/m	0.5 W/m	0.5 W/m	0.10	0.21
	BS	70	2.5 μW/cm^2	0.1 μW/cm^2	0.1 μW/cm^2	0.107	
Darnytsia	KTTs	1	3 W/m	0.4 W/m	0.4 W/m	0.10	0.161
	BS	80	2.5 μW/cm^2	0.1 μW/cm^2	0.1 μW/cm^2	0.061	

Table 4.8 Consumption of certain types of energy materials and oil refined products in 2013 for the Kyiv city

Energy materials and petroleum products	Amount	Tons of conventional fuel [t.c.f.]
Coal (t)	464.6	406,814
Natural gas (million m^3)	4072.3	4,642,422
Gasoline (million m^3)	458.6	683,314
Diesel fuel (million m^3)	361.3	523,885
Heavy fuel oil (t)	1.1	1507
Briquettes, pellets and similar solid fuels from coal (t)	0.2	72
Firewood (10^3 m^3)	30.2	8033
Fuel briquettes and pellets from wood and other natural raw materials (t)	0.1	36
Total	5388.4	6,266,083

The EMR emergy is the lowest (1.82×10^{11} to 1.72×10^{12} seJ), despite their usual energy indicators, which indicates the insignificant potential of this component for the functioning of the urban ecosystem.

The ratio of consumption indicators in the districts of Kyiv in the total component of the green areas is (−2.42)–(+0.39), which is quite low. Only in three districts of the city the attitude is positive: Holosiiv, Darnytsia, Svyatoshyn, but this effect is mainly achievable due to forestry enterprises located within these districts. The lowest among the districts of the city were Shevchenko, Dnipro, Pechersk, Obolon.

This means that the functioning of the city's SLAA system is quite energy-intensive. An increase in the influence of chemical and physical factors and a decrease in the area of green areas leads to a decrease and disruption of the energy balance of the planet, and the biosphere in general.

Table 4.9 Emerging balance of green areas in Kyiv city's districts

Anthropogenic and natural components of balance	Indicators of energy and resources (J)	Transformation index (seJ/J)	Emergy (seJ)	Balance
Holosiiv				
Noise load	3.53×10^{10}	1268	4.47×10^{13}	
EMR	1.36×10^{9}	1268	1.72×10^{12}	
Consumed energy [t_{fe}]	1.56×10^{13}	40,000	6.24×10^{17}	
Green area	1.13×10^{15}	1413	1.60×10^{18}	+0.39
Darnytsia				
Noise load	4.20×10^{10}	1268	5.32×10^{13}	
EMR	1.83×10^{8}	1268	2.32×10^{11}	
Consumed energy [t_{fe}]	2.11×10^{13}	40,000	8.45×10^{17}	
Green area	2.85×10^{15}	1413	4.02×10^{18}	+0.21
Desna				
Noise load	1.17×10^{11}	1268	1.49×10^{14}	
EMR	1.87×10^{8}	1268	2.37×10^{11}	
Consumed energy [t_{fe}]	2.39×10^{13}	40,000	9.58×10^{17}	
Green area	3.30×10^{14}	1413	4.66×10^{17}	−2.05
Dnipro				
Noise load	1.67×10^{10}	1268	2.12×10^{13}	
EMR	1.43×10^{8}	1268	1.82×10^{11}	
Consumed energy [t_{fe}]	2.28×10^{13}	40,000	9.12×10^{17}	
Green area	2.71×10^{14}	1413	3.82×10^{17}	−2.38
Obolon				
Noise load	1.10×10^{11}	1268	1.39×10^{14}	
EMR	1.07×10^{9}	1268	1.35×10^{12}	
Consumed energy [t_{fe}]	2.07×10^{13}	40,000	8.29×10^{17}	
Green area	2.48×10^{14}	1413	3.51×10^{17}	−2.36
Pechersk				
Noise load	6.19×10^{9}	1268	7.84×10^{12}	
EMR	9.23×10^{8}	1268	1.17×10^{12}	
Consumed energy [t_{fe}]	9.51×10^{12}	40,000	3.81×10^{17}	
Green area	1.14×10^{14}	1413	1.61×10^{17}	−2.37
Podil				
Noise load	4.27×10^{9}	1268	5.42×10^{12}	
EMR	6.10×10^{8}	1268	7.74×10^{11}	
Consumed energy [t_{fe}]	1.27×10^{13}	40,000	5.10×10^{17}	
Green area	2.77×10^{14}	1413	3.91×10^{17}	−1.30
Svyatoshyn				
Noise load	6.36×10^{10}	1268	8.06×10^{13}	
EMR	1.09×10^{9}	1268	1.38×10^{12}	

(continued)

Table 4.9 (continued)

Anthropogenic and natural components of balance	Indicators of energy and resources (J)	Transformation index (seJ/J)	Emergy (seJ)	Balance
Consumed energy [t_{fe}]	2.20×10^{13}	40,000	8.81×10^{17}	
Green area	2.24×10^{15}	1413	3.17×10^{18}	+0.28
Solom'yansk				
Noise load	1.59×10^{10}	1268	2.02×10^{13}	
EMR	8.87×10^{8}	1268	1.12×10^{12}	
Consumed energy [t_{fe}]	2.20×10^{13}	40,000	8.81×10^{17}	
Green area	4.66×10^{14}	1413	6.59×10^{17}	−1.40
Shevchenko				
Noise load	8.52×10^{9}	1268	1.08×10^{13}	
EMR	1.21×10^{9}	1268	1.53×10^{12}	
Consumed energy [t_{fe}]	1.51×10^{13}	40,000	6.03×10^{17}	
Green area	1.76×10^{14}	1413	2.49×10^{17}	−2.42

So, as energy transforms from natural to anthropogenic blocks of the urban ecosystem, the importance of emergy grows due to the work that a person performs to create goods, services and information resources.

4.3 Analysis of the Relationship Between the Green Area and the Surface Layer of Atmospheric Air

One of the statistical patterns in observing a pair of indicators X and Y is their dependence [24, 25].

The interaction of the resulting indicator (Y) with factorial features (X_1, X_2, …, X_n) is traditionally described by the linear multivariate regression equation, which is determined by the formula:

$$y = a_0 + a_1x_1 + a_2x_2 + a_3x_3 + \cdots + a_nx_n,$$

where y—calculated values of the effective characteristic; x_1, x_2,…—factor signs; a_0, a_1, a_2—parameters of the equation.

The state of the energy balance was chosen as the resulting indicator (Y), and the following indicators were chosen as factor values: X_1—SLAA hazard index; X_2—state of green spaces. The summary data for the correlation and regression analysis of the influence of factor values on the volume of sold innovative products are presented in Table 4.10.

As a result of solving the set goal of 8-factor correlation-regression analysis, an ecological-mathematical model was built, which has the following form:

Table 4.10 Summary data for correlation and regression analysis of the influence of factor values X_1, X_2 on the change in the energy balance Y

District	Y	X_1	X_2
Holosiiv	0.39	172	7578.85
Darnytsia	0.21	190	19,077.17
Desna	−2.05	209	2209.81
Dnipro	−2.38	174	1812.84
Obolon	−2.36	212	1661.71
Pechersk	−2.37	200	761.61
Podil	−1.3	225	15,006.92
Svyatoshyn	0.28	206	3125.23
Solom'yansk	−1.4	231	1180.73
Shevchenko	−2.42	209	2209.81
Total	−13.4	2028	54,624.68
Mean	−1.34	202.8	5462.468

$$y = 0.4871 + 0.0095 \cdot x_1 - 0.0001 \cdot x_2.$$

In the process of identifying the correlation and regression links between the effective indicator and factor values, in addition to building an economic and mathematical model, the multiple regression coefficient, the coefficient of determination, the standard error, the Student's t-test were calculated, which are shown in Tables 4.11, 4.12 and 4.13.

The multiple regression coefficient R = 0.925 indicates a very close relationship between the effective indicator and the factor values. Regarding the value of the coefficient of determination R^2 of the obtained correlation-regression model $R^2 = 0.855$, the dependence of the volume of sold innovative products by 85.5% is due to the selected factor values.

The remaining 14.6% is due to other factors that affect the volume of sold innovative products, but are not included in the regression model.

Considering the high values of the coefficients of multiple regression and determination, this relationship is quite natural. The indicator of variance, significance F and F-statistics indicator show a sufficient level of reliability of the assessment results.

Graphical interpretation of the calculated data for each indicator is shown in Fig. 4.1a, b.

Table 4.11 Regression statistics

Multiple regression coefficient R	0.925008441
Determination coefficient R^2	0.855640617
Normalized determination coefficient R^2	0.797896864
Standard error	0.40956714

Table 4.12 Indicators characterizing the reliability of the regression model

	DF	SS	MS	V	V significance
Regression	2	4.971261288	2.485630644	14.81789	0.007917946
Balance	5	0.838726212	0.167745242	–	–
Total	7	5.8099875	–	–	–

Table 4.13 Correlation coefficients

Parameters	Coefficients	Standard error	Student's criterion t	P-value
Y-crossing	0.487109014	1.481797119	0.328728547	0.755682
X_1	0.009526876	0.007293937	1.306136414	0.248363
X_2	−0.000172962	3.1996×10^{-5}	−5.40573251	0.002929

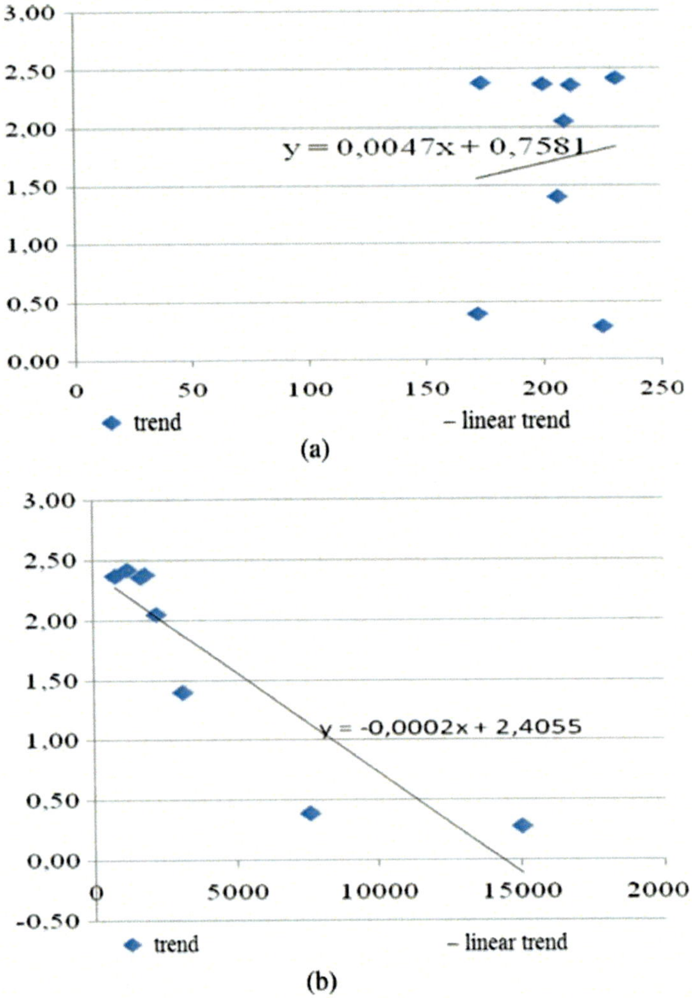

Fig. 4.1 Estimated data for each indicator: **a** pollution state change graph (X_1); **b** graph of changes in the green areas (X_2)

4.4 Conclusions

1. The results of the analysis and assessment of the network and the state of green areas of public use allow to determine the actual indicators of security in the city (16.5 m^2/person) and administrative districts.
2. According to the research results, the actual provision of green areas of public use in Darnytsia, Svyatoshyn and Solom'yansk districts is lower than the standardized (according to SBS 360-92 "Urban planning. Planning and development of urban and rural settlements", the area of green zones of general use for cities with a population of more than 1 million people should be at least 10–11 m^2/person).
3. According to the research results, a table of average values of levels of equivalent noise levels of day and night period by districts of the Kyiv city was formed.
4. It was determined that for a balanced development of the city, it is necessary to increase the green component of the city by 2.04 times. This means that the functioning of the city's atmospheric air systems are quite energy intensive. An increase in the amount of the influence of chemical and physical factors and a decrease in the area of green spaces leads to a decrease and disruption of the energy balance of the planet.
5. As a result of solving the set goal of 8-factor correlation-regression analysis, an ecological-mathematical model was built. In the process of identifying correlations and regression links between the effective indicator and factor values, in addition to building an economic and mathematical model, the multiple regression coefficient, the coefficient of determination, the standard error, and the Student's t-criteria were calculated.

References

1. Goncharenko, I., Semenischenkov, Yu., Tsakalos, J.L., Mucina, L.: Thermophilous oak forests of the steppe and forest-steppe zones of Ukraine and Western Russia. Biologia **75**, 337–353 (2020). https://doi.org/10.1556/ABot.55.2013.1-2.11
2. Didukh, Y., Alioshkina, U.: Ecotope classification of Kyiv. Naukovi zapysky, Biologia and Ecol. **54**, 50–57 (2006). (in Ukrainian)
3. Morar, T., Radoslav, R., Spiridon, L.C., Pacurar, L.: Assessing pedestrian accessibility to green space using GIS. Transylvanian Rev. Adm. Sci. **42**, 116–139 (2014)
4. Colding, J., Lundberg, J., Folke, C.: Incorporating green-area user groups in urban ecosystem management. AMBIO: A J. Human Environ. **35**(5), 237–244 (2006). https://doi.org/10.1579/05-A-098R.1
5. Maer-Abih, K.M.: Uprising to protect nature: from the environment to the community (2004)
6. Dronova, O., Maruniak, E.: Changing the symbolic language of the urban landscape: post-socialist transformation in Kyiv. In: Brunn, S., Kehrein, R. (eds.) Handbook of the Changing World Language Map, pp. 2941–2972. Springer, Cham (2020). https://doi.org/10.1007/978-3-030-02438-3_117

7. Pleshkanovska, A.M.: Assessing the level of greening in a major city: subjective and objective evaluation on the example of the city of Kyiv. Bull. Geogr. Socio-economic Ser. **48**(48), 155–164 (2020). https://doi.org/10.2478/bog-2020-0019
8. State building standards 360-92.: Urban planning. Planning and construction of urban and rural settlements. Ministry of Construction and Architecture, Kyiv (1993)
9. Movchan, Y.I.: National ecological network of Ukraine: concept and implementation scenarios (2001)
10. Senetra, A., Krzywnicka, I., Mielke, M.: An analysis of the spatial distribution, influence and quality of urban green space—a case study of the Polish city of Tczew. Bull. Geogr. Socio-economic Ser. **42**(42), 129–149 (2018). https://doi.org/10.2478/bog-2018-0035
11. Ignatieva, M., Stewart, G.H., Meurk, C.: Planning and design of ecological networks in urban areas. Landscape Ecol. Eng. **7**, 17–25 (2011). https://doi.org/10.1007/s11355-010-0143-y
12. Odum, Yu.: Osnovy ekologii (Fundamental Ecology). Mir, Moscow (1975)
13. Lei, K., Zhou, S., Wang, Z.: Emergy synthesis and simulation for Macao. In: Ecological Emergy Accounting for a Limited System: General Principles and a Case Study of Macao. Springer Environmental Science and Engineering, pp. 67–85. Springer, Berlin, Heidelberg (2014). https://doi.org/10.1007/978-3-642-45170-6_3
14. Zhang, L.X., Ulgiati, S., Yang, Z.F., Chen, B.: Emergy evaluation and economic analysis of three wetland fish farming systems in Nansi Lake area. China. J. Environ. Manag. **92**(3), 683–694 (2011). https://doi.org/10.1016/j.jenvman.2010.10.005
15. Ulgiati, S., Brown, M.T.: Emergy evaluations and environmental loading of alternative electricity production systems. J. Clean. Prod. **10**, 335–348 (2001)
16. Brown, M.T., Ulgiati, S.: Assessing the global environmental sources driving the geobiosphere: a revised emergy baseline. Ecol. Model. **339**, 126–132 (2016). https://doi.org/10.1016/j.ecolmodel.2016.03.017
17. Osmond C.B., Björkman O., Anderson D.J.: Physiological processes in plant ecology: the structure for a synthesis. In: Physiological Processes in Plant Ecology. Ecological Studies (Analysis and Synthesis), vol. 36, pp. 1–11. Springer, Berlin, Heidelberg (1980). https://doi.org/10.1007/978-3-642-67637-6_1
18. Gulevets, D.V.: Modeling the state of development of the urban environment in the context of environmental safety. In: Environmental security of the state—2015, pp. 187–188 (2015). (in Ukrainian)
19. Dumansky, V.Y.: Hygienic assessment of the electromagnetic situation and scientific substantiation of its safety requirements in modern settlements of Ukraine. The dissertation on competition of a scientific degree of the doctor of medical sciences on a specialty, 14 (2009). (in Ukrainian)
20. Zaporozhets, A., Kovtun, S., Dekusha, O.: System for monitoring the technical state of heating networks based on UAVs. In: Shakhovska, N., Medykovskyy, M. (eds.) Advances in Intelligent Systems and Computing IV, CCSIT 2019, vol. 1080, pp. 935–950. Springer, Cham (2020). https://doi.org/10.1007/978-3-030-33695-0_61
21. Zaporozhets, A.: Overview of quadrocopters for energy and ecological monitoring. In: Babak V., Isaienko V., Zaporozhets, A. (eds.) Systems, Decision and Control in Energy I. Studies in Systems, Decision and Control, vol. 298, pp. 15–36. Springer, Cham (2020). https://doi.org/10.1007/978-3-030-48583-2_2
22. Berezhnyi, E.O., Havrylenko, V.M., Hryhorovskiy, V.V., Evtushenko, L.V., Zaporozhets, O. I., Movchan, Ya.I., Pashkevych, M.O., Tron, O.Y., Hulevets, D.V.: Energy efficiency of Ukraine: current challenges. Bull. NAS of Ukraine **7**, 61–69 (2013). (in Ukrainian)
23. Zaporozhets, O., Movchan, Y., Gavrylenko, V., Gulevets, D.: Energy efficiency of Ukraine. Proc. Natl. Aviat. Univ. **48**(3), 176–181 (2011)

24. Zaporozhets, A., Eremenko, V., Redko, O.: Metrological assessment of the indirect method of measuring the concentration of oxygen in the air. In: 2019 IEEE 8th International Conference on Advanced Optoelectronics and Lasers (CAOL), pp. 640–643. Sozopol, Bulgaria (2019). https://doi.org/10.1109/caol46282.2019.9019506
25. Babak V.P., Babak S.V., Myslovych M.V., Zaporozhets A.O., Zvaritch V.M.: Methods and models for information data analysis. In: Diagnostic Systems for Energy Equipments. Studies in Systems, Decision and Control, vol. 281, pp. 23–70. Springer, Cham (2020). https://doi.org/10.1007/978-3-030-44443-3_2

Chapter 5
Improving the Environmental Security of Urban Areas on the Basis of GIS and Web Technology

Abstract In this chapter, the concept of a scheme for informing the population about the state of pollution of the surface layer of atmospheric air is proposed to improve the environmental safety of an urbanized area in real time. The main goal of creating a scheme for informing the population about the state of pollution of the surface layer of atmospheric air is to organize the collection, processing, storage and issuance of information, which will provide information support for decision-making aimed at ensuring the ecological and hygienic well-being of an urbanized territory. GIS maps will form the basis for analyzing the possible impact on the population of various factors of environmental pollution. In order for the results of mathematical modeling of environmental monitoring data to contribute to decision-making in the current situation, the results of monitoring measurements should be easily transferred to the GIS and, conversely, the data from the GIS should be recognized and used to inform the population in real time. With a successful and correct combination of mathematical models, GIS and web technologies, it is possible to achieve the maximum effect from the results of monitoring the state of the surface layer of atmospheric air in an urbanized area.

Keywords Population awareness · GIS maps · Web technologies · Mathematical modeling · Urbanized area

5.1 Environmental Monitoring of an Urbanized Area Based on Geoinformation Technologies

As indicated in the concept of the state program for environmental monitoring [1, 2], the functioning of the state environmental monitoring system of Ukraine, including the SLAA, is currently unsatisfactory. At present, the overwhelming majority of scientific works, which propose certain approaches to modernizing the existing atmospheric monitoring system in Ukraine, are either too conceptual in nature, or relate to the engineering component of the problem, that is, the development of software and hardware for measuring equipment. The above conditions

D. Gulevets et al., *Environmental Control for Ensuring Cities Safety*,
Lecture Notes in Intelligent Transportation and Infrastructure,
https://doi.org/10.1007/978-3-030-66710-8_5

require a fundamentally new approach to improving the information technology system and structures for receiving, processing and presenting data on the quality of SLAA [3–5].

The implementation of ecological information systems and software products based on modern geoinformation technologies and web technologies for the operational solution of monitoring tasks of SLAA [6–8] is becoming urgent.

In the study of SLAA pollution, of interest are not only the levels of pollution in the city (district) as a whole, which can be obtained from government agencies involved in the environmental monitoring system, but also from individual sources. Enterprises can make a big contribution to the air pollution of the city (district). During organizing a network for monitoring the level of air pollution in cities, it should be borne in mind that some low-lying (vehicles) and even small (chimneys of residential buildings) sources can affect the local level of pollution more significantly than high-located sources (discharge from high pipes).

Having established the main factors that can affect the level of environmental safety of the SLAA, it should be gone directly to the tool with which the specified level is maintained. Environmental monitoring provides for control over the environmental state, assessment of the power of environmental impact and forecasting the consequences of it for human health and the environment [9–12].

From a methodological point of view, "critical" objects, which are permanent sources of air pollution, should be monitored in such a way that their emissions are controlled by automated means and provide an opportunity to receive data on the state of the SLAA for the population in real time. Automated control systems must necessarily become basic in cities and industrial centers with constant and intense traffic flows. The revealed deviations from the established norms are subject to analysis and should form the basis of economic compensation for the caused environmental damage.

Given the extreme urgency in Ukraine of solving the problems of improving the air basin, the expediency of a radical change in the situation for the better does not raise any doubts [1, 2].

Much attention is paid to the issues of building traditional environmental monitoring systems. They usually include:

- widespread use of aerospace information as part of other remote sensing devices, contact measurements and biotesting;
- taking into account the whole complex of medical and hygienic assessments;
- a set of observations of the dynamics of socio-economic indicators of society, etc.

In this regard, it is necessary to note the effectiveness of using the specified class of GIS as a basis for solving problems. With the help of GIS technology, it is possible to effectively bring all the monitoring information in space-time coordinates. Exactly this opens the way for creating conditions for the effective and prompt determination of directions and means of liquidating the negative consequences of emergencies and solving key issues of environmental safety.

To implement this monitoring structure, a well-functioning material base of the block for developing environmental solutions is needed, which would consist of the following means:

- receiving and registering information (including remote sensing data obtained during contact measurements, biotesting, as well as relevant statistical data and results of sociological surveys, etc.);
- provision of systems for preliminary analysis and distribution of information, its rejection and transmission to the input of the interbranch processing system or other centers;
- provision of cross-industry processing, which brings information to a state acceptable to a wide range of users;
- databanks for collecting information that can be used to develop appropriate decisions on regional, state and interstate issues [13–15].

The entire environmental complex of problems is associated with a number of tasks based on the examination of already existing objects (identifying the degree of criticality and establishing the most risky links) and those that are planned to be built (in accordance with the established criticality standards). In addition, it is necessary to provide for work on finding a balance between the ecological capacity and anthropogenic loads of UA with a difficult ecological state and the search for the most optimal ways of implementation [16–18].

The main goal of creating an environmental monitoring system of SLAA in real time is to organize, on the basis of geoinformation technologies, an intersectoral and hierarchical system for collecting, processing, storing and issuing information, which provides continuous diagnostics of the state of the environment, information support for decision-making aimed at ensuring the ecological and hygienic well-being of the population [19–21].

Environmental monitoring of SLLA of UA based on GIS technologies is:

- system for observing changes in chemical and physical factors of influence based on the natural environment of SLAA (chemical, noise, electromagnetic);
- development of measures to improve the natural environment of the city and minimize environmental risk.

The tasks of environmental monitoring of UA based on GIS technologies are:

- control of the content of various chemical pollutants and physical factors (noise, electromagnetic) impact on the city environment;
- identification of changes in the content of pollutants and technogenic load in the SLAA;
- calculation of levels of environmental risk and forecasting the occurrence of environmentally-related consequences for the population;
- identification of priority technogenic, urban planning, meteorological factors of pollution of the natural environment of the city;

- detailed analysis of the register of emissions of industrial enterprises, especially the analysis of emissions of specific little-studied factors of noise, electromagnetic, pollution of various environments;
- information support of environmental protection services, environmental control and health authorities with the results of exploratory research on health risk assessment and spatial distribution of risk zones.

Our proposed scheme of ecological monitoring of an urbanized territory based on geoinformation and web technologies is implemented on the example of Kyiv city.

A necessary and sufficient condition for the functioning of the proposed integrated software solution for assessing environmental risk and its reflection in the GIS is the sequential solution of stages, including the ecological zoning of territories according to the I_{haa} excess indicator, the interpretation of zones according to the I_{haa} excess indicator in the environmental risk zones, the mechanism for displaying environmental risk in a multi-level matrix GIS.

The construction of an integrated software solution for assessing environmental risk and its reflection in GIS is based on the integration of three heterogeneous operational blocks [22]. The first block includes a program for calculating air pollution with zoning of territories according to the SLAA hazard index. The second block includes the environmental risk assessment described in this book for the zoning of the territory according to the risk indicator. The third block includes GIS for displaying the state of SLAA pollution and environmental risk (Fig. 5.1).

Figure 5.2 shows the algorithmic form of an integrated software solution for assessing the state of pollution of the SLAA/environmental risk, and its reflection in the GIS. The environmental risk assessment module conducts zoning of the territory by the risk indicator using the formulas.

The creation of this database is carried out using the MS Excel software package, which today is the most optimal for the implementation of the tasks. The database contains information on the content of SLAA pollutants (including noise and electromagnetic pollution) according to the PMP of UA. For PMP of SLAA, next microclimatic indicators are indicated—air temperature, relative humidity, atmospheric pressure, wind direction and speed.

Within the framework of the DB, the values of the SLAA hazard index are calculated, depending on the atmospheric pollution index from chemical exposure, and the index of pollution from physical exposure, the magnitude of environmental risks for each sampling point, and average indicators for districts and the city as a whole.

Within the framework of the geographic and cartographic block, it is planned to annually update the cartographic material of the city, to apply on an electronic map new residential objects, construction projects begun, objects of environmental risk —gas stations, industrial facilities, highways, etc. In addition, an increase or decrease in the recreational area, green spaces, gardens, parks, etc. must be recorded on the electronic map.

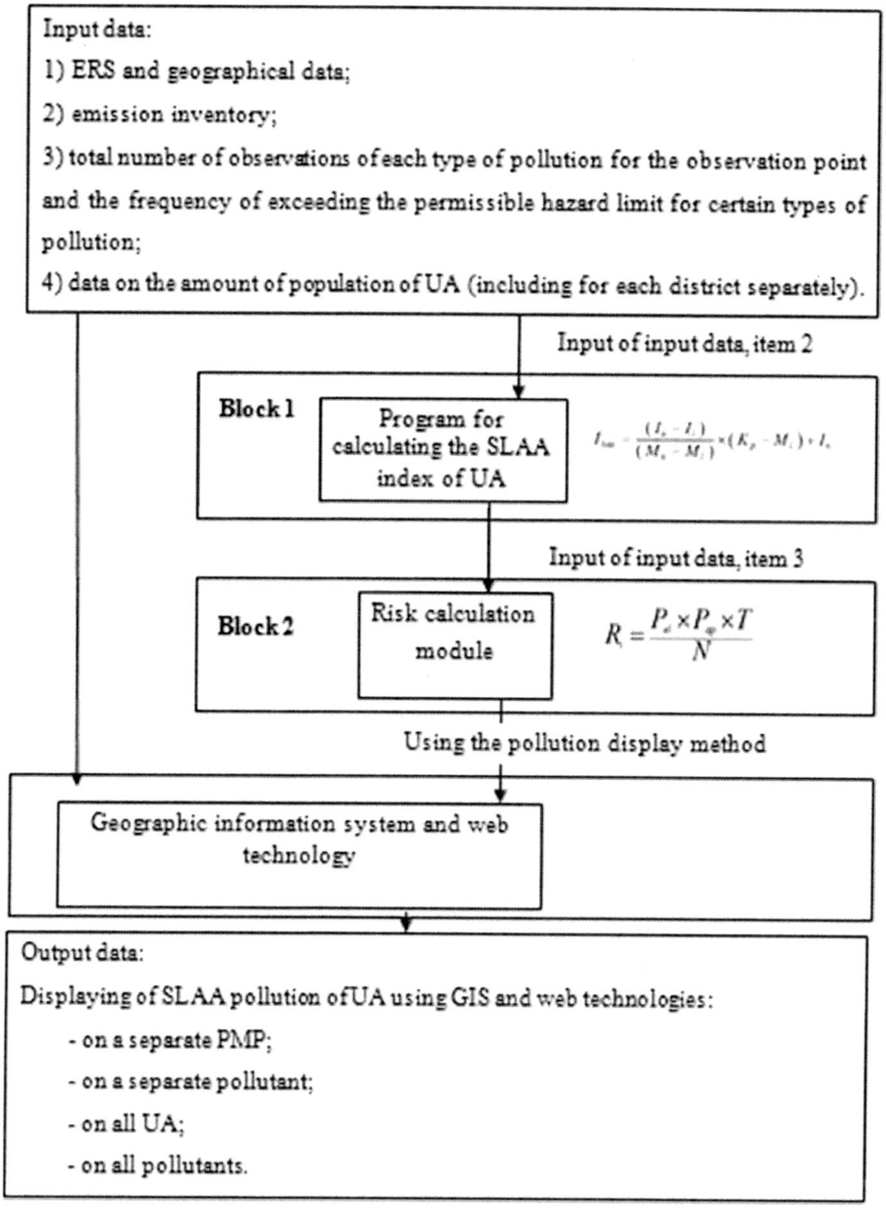

Fig. 5.1 Diagram of the structure of an integrated software solution for assessing environmental risk and its reflection in GIS

Within the framework of this block, it is necessary to carry out an annual collection of statistical information on environmental, meteorological and climatic data—air temperature, relative humidity, atmospheric pressure, wind speed and direction in various districts and quarters with multi-storey buildings in the city.

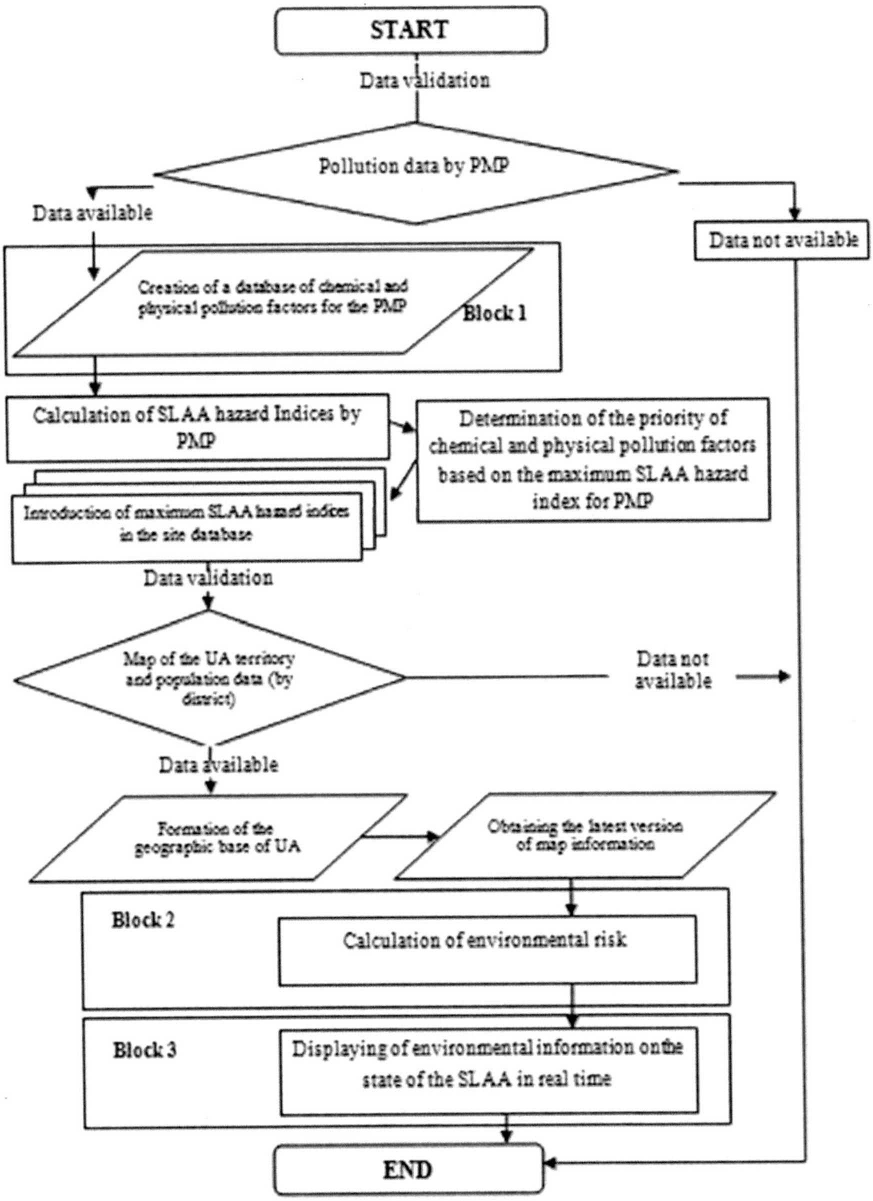

Fig. 5.2 Algorithm for using an integrated software solution for assessing environmental risk and its reflection in GIS

Thus, an automated database is the main component of the GIS with the help of which environmental monitoring is carried out.

In addition, it seems expedient to include an "Emergency block" in the GIS, within which, based on complaints from the population or information about industrial accidents at environmental risk facilities, an unscheduled analysis of SLAA pollution, which may be caused by urban environmental factors, is performed.

Within the framework of the assessment and analytical block, it is possible to assess the factors of technogenic load on the formation of environmental risk zones, assess urban planning factors in the formation of fields of technogenic pollution of natural environments, assess the incidence of the population, considered as a "recall of urban ecosystems" on the formation of fields of environmental and geochemical pollution of natural environments [16, 23].

Thus, a scheme of ecological and geochemical monitoring of UA has been developed based on geoinformation and web technologies, which can be implemented in most industrial cities of Ukraine.

5.2 Web-Based Real-Time Displaying of Environmental Information

The development of DB technology began with the need to separate data used for calculation operations and data used for analytical purposes. DBs provide the capabilities best suited for reporting.

The main goal of creating a SLAA environmental monitoring system is to organize, on the basis of geoinformation technologies, an intersectoral and hierarchical system for collecting, processing, storing and issuing information, which provides continuous diagnostics of the state of the environment and information support for decision-making aimed at ensuring environmental and hygienic well-being.

In order for the results of mathematical modeling of environmental and geochemical monitoring data to facilitate decision-making in the current situation, the results of monitoring measurements should be easily transferred to the GIS and the website, and vice versa, the data from the GIS should be recognized and used in calculations during constructing mathematical models [17, 24]. With a successful and correct combination of mathematical models and GIS, it is possible to achieve the maximum effect from the results of monitoring and modeling and expand the scope of GIS (Fig. 5.3).

The project of an integrated software solution is divided into three logical parts:

- editing of pollution types and collection point;
- adding input data;
- display of statistical data.

In terms of editing information collection points, it can be used the software product "Google Map" and "APIn" (Application Programming Interface). These software products provide advanced functionality for finding the address of the PMP, collecting information, displaying it on a map and the possibility of manual correction of coordinates.

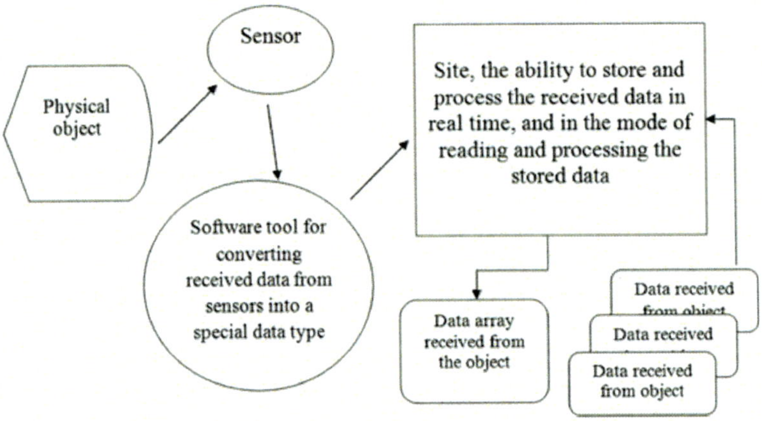

Fig. 5.3 Ecological monitoring scheme for an UA

In terms of editing information collection points, it can be used the software product "Google Map" and "APIn" (Application Programming Interface) These software products provide advanced functionality for finding the address of the PMP, collecting information, displaying it on a map and the possibility of manual correction of coordinates.

It is possible to add input data separately for each PMP by filling out a form with the date and types of pollution, as well as by loading a database with a clearly defined template.

In the future, in order to automate the process of collecting information, it is necessary to develop an application programming interface for automatically receiving data from the PMP regarding the SLAA pollution. Figure 5.4 shows the structure of the web portal for determining the state of the SLLA of UA.

The database is divided into linked tables. For implementation, the table structures, discussed below, were created.

Table "eco_types"—a list of pollution types, with the ability to add new and edit old types. The link with the "eco_type_group" table is carried out by the "GroupID" field.

This table contains information about the type of measurements, name, maximum allowable parameter, calculation measure (mg/m^3, dB), and linked to the "eco_type_group" table to determine the measurement group (Fig. 5.5).

The structure of the "eco_types" table:

- "Id"—serial number;
- "Name"—the name of the pollution;
- "GroupID"—pollution group number;
- "Param"—the maximum permissible level of pollution for this type;
- "Symbol"—units of measurement for this type (mg/m^3, dB).

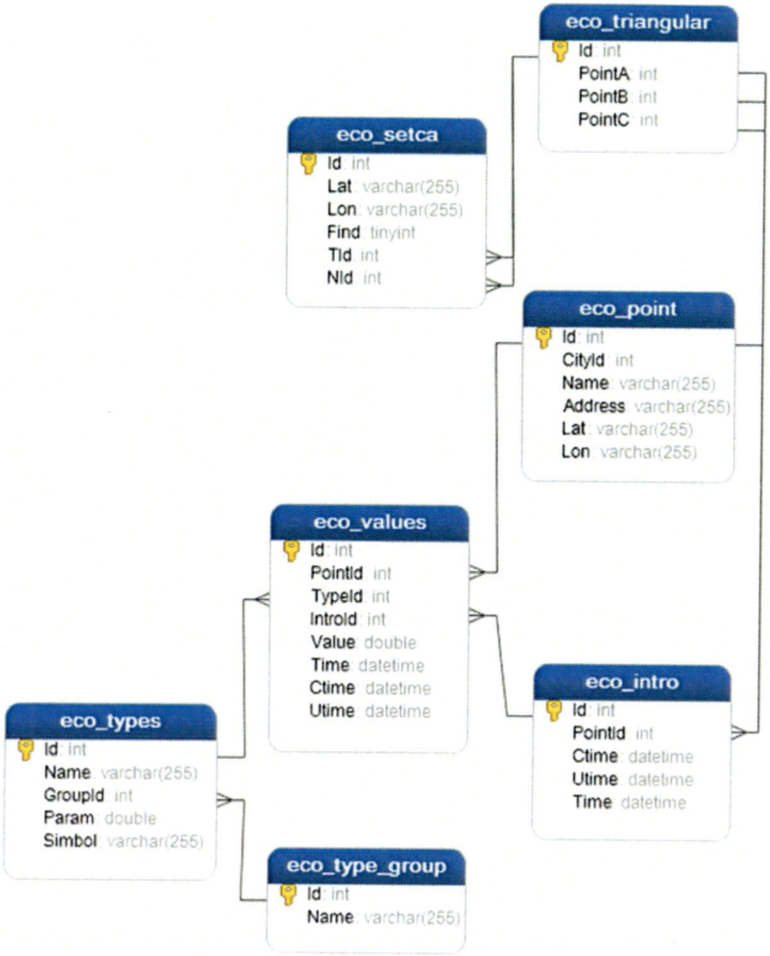

Fig. 5.4 Web portal structure

The grid of intermediate measurement points stores the coordinates of the PMP point and other possible PMP or NMP to increase the calculations of the state of pollution of the SLAA of UA. Data are entered into the "eco_setca" table automatically. Then intermediate points are automatically created and stored (Fig. 5.6).

The structure of the "eco_setca" table:

- "Id"—serial number;
- "Lat"—latitude coordinate for displaying on the map and further calculations;
- "Lon"—longitude coordinate;
- "TId"—the number of the triangle, which includes the PMP;
- "NId"—number of the nearest PMP.

Id	Name	GroupId	Param	Simbol
19	Хлористий водень	1	0.2	мг/м3
17	Фенол	1	0.003	мг/м3
16	Сірководень	1	0.008	мг/м3
13	Оксид вуглецю	1	3	мг/м3
18	Фористий водень	1	0.005	мг/м3
15	Оксид азоту	1	0.06	мг/м3
14	Діоксид азоту	1	0.04	мг/м3
12	Діоксид сірки	1	0.05	мг/м3
11	Завислі речовини (пил)	1	0.15	мг/м3
20	Аміак	1	0.04	мг/м3
21	Формальдегід	1	0.003	мг/м3
22	Еквівалентний рівень шуму	2	(Null)	дБ

Fig. 5.5 Structure of the "eco_types" table

The "eco_setca" table is a list of points where pollution indicators will be calculated on the basis of which a pollution map will be built. This table is linked to the "eco_triangular" table by the "TId" field and "eco_point" by the table by the "NId" field.

The "eco_values" table is a collection and display of pollution data for the SPS UT. This table is linked to the "eco_intro" table by the "Introld" field and to the "eco_types" table by the "TypeId" field.

The "eco_values" table is a collection and display of pollution data for the PMP of UA. This table is linked to the "eco_intro" table by the "Introld" field and to the "eco_types" table by the "TypeId" field.

The table "eco_values" stores the following data: the number of the measurement point (PMP or NMP) where the measurements were made, the number of the measurement, the number of the input data point, the measurement values and the date the measurement was entered into the system (Fig. 5.7).

The structure of the "eco_values" table:

- "Id"—serial number of the operation;
- "PointId"—number of the information collection point;
- "TypeId"—number of the type of pollution (chemical or physical);
- "Value"—value of the SLAA pollution level;
- "Time"—date and time of collection of statistical data;
- "Ctime"—date and time of the record creation;
- "Utime"—date and time when the entry was modified;
- "Introld"—input data number.

Id	Lat	Lon	Find	TId	NId
498	50.446452638756	30.452016816739	1	11	(Null)
499	50.401486558461	30.536719830246	1	13	(Null)
500	50.401486558461	30.522608022158	1	5	(Null)
501	50.473432286933	30.452000713063	1	2	(Null)
502	50.464439070874	30.452000713063	1	10	(Null)
503	50.401486558461	30.564959529594	1	(Null)	8
504	50.401486558461	30.550847721506	1	6	(Null)
505	50.491418719051	30.593258274409	1	(Null)	3
506	50.473432286933	30.593258274409	1	(Null)	3
507	50.482425502992	30.593258274409	1	(Null)	3
508	50.455445854815	30.5932528983	1	21	(Null)
509	50.464439070874	30.5932528983	1	9	(Null)
510	50.437459422697	30.593242154304	1	14	(Null)
511	50.446452638756	30.593242154304	1	8	(Null)
512	50.419472990579	30.593226050628	1	6	(Null)
513	50.428466206638	30.593226050628	1	6	(Null)
514	50.491418719051	30.451979225067	1	(Null)	16
515	50.482425502992	30.451979225067	1	2	(Null)
516	50.401486558461	30.593204595466	1	(Null)	8

Fig. 5.6 Structure of the "eco_setca" table

The "eco_triangular" table is necessary to preserve the structure of areas and is divided into triangles for further modeling of pollution using the Delaunay triangulation method. The "eco_triangular" table stores connected triangles of points for calculations, and is linked to the "eco_point" table (Fig. 5.8). The links of this table are carried out by the fields PointA, PointB and PointC of the table "eco_point".

The structure of the "eco_triangular" table:

- "Id"—serial number;
- "PointA"—first point of information collection;
- "PointB"—second point of information collection;
- "PointC"—third point of information collection.

The "eco_point" table stores the coordinates of points in a rectangular coordinate system from different measurement points (PMP or NMP) from which measurements of the environmental state of the SLAA are received. The "eco_point" table is linked by the "CityId" field with an additional UA table for the possibility of further adding information from other cities (Fig. 5.9).

Id	PointId	TypeId	Value	Time	Ctime	Utime	IntroId
1	1	11	0.0999	2015-01-01 00:00:00	2016-03-31 11:56:45	2016-03-31 11:56:45	1
2	1	12	0.0156	2015-01-01 00:00:00	2016-03-31 11:56:45	2016-03-31 11:56:45	1
3	1	13	0.1399	2015-01-01 00:00:00	2016-03-31 11:56:45	2016-03-31 11:56:45	1
4	1	14	0.1014	2015-01-01 00:00:00	2016-03-31 11:56:45	2016-03-31 11:56:45	1
5	1	17	0.0045	2015-01-01 00:00:00	2016-03-31 11:56:45	2016-03-31 11:56:45	1
6	1	20	0.0053	2015-01-01 00:00:00	2016-03-31 11:56:45	2016-03-31 11:56:45	1
7	1	21	0.0051	2015-01-01 00:00:00	2016-03-31 11:56:45	2016-03-31 11:56:45	1
8	1	11	0.0999	2015-02-01 00:00:00	2016-03-31 11:56:45	2016-03-31 11:56:45	2
9	1	12	0.0143	2015-02-01 00:00:00	2016-03-31 11:56:45	2016-03-31 11:56:45	2
10	1	13	0.2291	2015-02-01 00:00:00	2016-03-31 11:56:45	2016-03-31 11:56:45	2
11	1	14	0.1146	2015-02-01 00:00:00	2016-03-31 11:56:45	2016-03-31 11:56:45	2
12	1	17	0.0045	2015-02-01 00:00:00	2016-03-31 11:56:45	2016-03-31 11:56:45	2
13	1	20	0.0043	2015-02-01 00:00:00	2016-03-31 11:56:45	2016-03-31 11:56:45	2
14	1	21	0.0049	2015-02-01 00:00:00	2016-03-31 11:56:45	2016-03-31 11:56:45	2
15	1	11	0.0999	2015-03-01 00:00:00	2016-03-31 11:56:45	2016-03-31 11:56:45	3
16	1	12	0.0124	2015-03-01 00:00:00	2016-03-31 11:56:45	2016-03-31 11:56:45	3
17	1	13	0.2599	2015-03-01 00:00:00	2016-03-31 11:56:45	2016-03-31 11:56:45	3
18	1	14	0.0941	2015-03-01 00:00:00	2016-03-31 11:56:45	2016-03-31 11:56:45	3
19	1	17	0.0021	2015-03-01 00:00:00	2016-03-31 11:56:45	2016-03-31 11:56:45	3
20	1	20	0.0049	2015-03-01 00:00:00	2016-03-31 11:56:45	2016-03-31 11:56:45	3
21	1	21	0.0036	2015-03-01 00:00:00	2016-03-31 11:56:45	2016-03-31 11:56:45	3
22	1	11	0.1019	2015-04-01 00:00:00	2016-03-31 11:56:45	2016-03-31 11:56:45	4
23	1	12	0.0102	2015-04-01 00:00:00	2016-03-31 11:56:45	2016-03-31 11:56:45	4
24	1	13	0.4807	2015-04-01 00:00:00	2016-03-31 11:56:45	2016-03-31 11:56:45	4
25	1	14	0.0907	2015-04-01 00:00:00	2016-03-31 11:56:45	2016-03-31 11:56:45	4
26	1	17	0.0013	2015-04-01 00:00:00	2016-03-31 11:56:45	2016-03-31 11:56:45	4
27	1	20	0.0046	2015-04-01 00:00:00	2016-03-31 11:56:45	2016-03-31 11:56:45	4
28	1	21	0.0067	2015-04-01 00:00:00	2016-03-31 11:56:45	2016-03-31 11:56:45	4
29	1	11	0.0978	2015-05-01 00:00:00	2016-03-31 11:56:45	2016-03-31 11:56:45	5
30	1	12	0.0098	2015-05-01 00:00:00	2016-03-31 11:56:45	2016-03-31 11:56:45	5
31	1	13	0.5652	2015-05-01 00:00:00	2016-03-31 11:56:45	2016-03-31 11:56:45	5
32	1	14	0.1154	2015-05-01 00:00:00	2016-03-31 11:56:45	2016-03-31 11:56:45	5
33	1	17	0.001	2015-05-01 00:00:00	2016-03-31 11:56:45	2016-03-31 11:56:45	5

Fig. 5.7 Structure of the "eco_values" table

The "eco_point" table also stores information about the city, address and coordinates. Each city is represented by a corresponding number. To add cities, it is possible to additionally create a table of cities and link new measurement points for each city.

The structure of the "eco_point" table:

- "Id"—serial number;
- "CityId"—city number;
- "Name"—name of the information collection point;
- "Address"—address of the collection point;
- "Lat"—latitude coordinate for displaying on the map and subsequent calculations (rectangular coordinate system);
- "Lon"—longitude coordinate (rectangular coordinate system).

The "eco_intro" table registers the history of the input data from the measurement points and the PPS or PSS. This table is linked to the "eco_point" table and the "eco_point" table by the "PointId" field.

Fig. 5.8 Structure of the "Eco_triangular" table

Id	PointA	PointB	PointC
1	14	16	10
2	16	11	2
3	11	1	2
4	1	12	15
5	12	5	15
6	5	9	8
7	9	4	3
8	3	9	13
9	3	14	10
10	16	2	10
11	1	2	6
12	1	6	15
13	5	15	8
14	8	9	13
15	2	6	10
16	10	7	6
17	6	7	15
18	7	15	8
19	7	13	8
20	13	7	10
21	13	10	3
22	5	9	4

The "eco_intro" table stores the measuring point number, date and time of addition. This structure can be extended with the field of an individual key to protect against data corruption during using the "APIn" (Fig. 5.10).

The structure of the "eco_intro" table:

- "Id"—serial number;
- "CTime"—date and time of the record creation;
- "Utime"—date and time when the entry was modified;
- "Time"—date and time of collection of statistical data;
- "PointId"—number of the information collection point.

In the reflection of statistical data, a map of the area is displayed with information collection points and information values of the degree of pollution, highlighted in certain colors, depending on the degree of pollution: green, orange, purple, red and brown in accordance with the safe level of pollution, moderate level of pollution, dangerous level for a sensitive population of pollution, a dangerous level of pollution, and a very dangerous level of pollution (Fig. 5.11).

Id	CityId	Name	Address	Lat	Lon
1	1	ПС №1	ул. Академика Стражеска,	50.43276634402626	30.414753282431093
2	1	ПС №2	ул. Довженка, 8	50.45780850000001	30.447640999999976
3	1	ПС №3	ул. Попудренко, 50	50.459384	30.63313949999997
4	1	ПС №4	ул. Лазо, 2	50.4468165	30.67104649999999
5	1	ПС №5	просп. Науки, 37	50.391982	30.533492000000024
6	1	ПС №6	площадь Победы	50.44713789999999	30.492637699999932
7	1	ПС №7	площадь Бессарабская	50.441978	30.521248000000014
8	1	МС №8	бул. Леси Украинки, 29	50.4205575	30.54741999999999
9	1	ПС №9	ул. Каунаская, 10а	50.434708	30.626756500000056
10	1	ПС №10	ул. Межигорская 56/60	50.472747	30.506304
11	1	ПС №11	просп. Победы 98/2	50.4586996	30.396787899999936
12	1	ПС №13	Експоцентр Украины	50.37987292879188	30.477672809509272
13	1	ПС №15	Гідропарк	50.443611	30.576944000000026
14	1	ПС №17	просп Оболонский, 14	50.505603	30.499262000000044
15	1	ПС №20	Московська площа	50.4064716	30.5186291
16	1	ПС №21	ул. Скляренка, 5	50.4933097	30.468692799999985

Fig. 5.9 Structure of the "eco_point" table

Id	Ctime	Utime	Time	PointId
20	2016-03-31 11:56:45	2016-03-31 11:56:45	2015-08-01 00:00:00	2
21	2016-03-31 11:56:45	2016-03-31 11:56:45	2015-09-01 00:00:00	2
22	2016-03-31 11:56:45	2016-03-31 11:56:45	2015-10-01 00:00:00	2
23	2016-03-31 11:56:45	2016-03-31 11:56:45	2015-11-01 00:00:00	2
24	2016-03-31 11:56:45	2016-03-31 11:56:45	2015-12-01 00:00:00	2
25	2016-03-31 11:56:45	2016-03-31 11:56:45	2015-01-01 00:00:00	3
26	2016-03-31 11:56:45	2016-03-31 11:56:45	2015-02-01 00:00:00	3
27	2016-03-31 11:56:45	2016-03-31 11:56:45	2015-03-01 00:00:00	3
28	2016-03-31 11:56:45	2016-03-31 11:56:45	2015-04-01 00:00:00	3
29	2016-03-31 11:56:45	2016-03-31 11:56:45	2015-05-01 00:00:00	3
30	2016-03-31 11:56:45	2016-03-31 11:56:45	2015-06-01 00:00:00	3
31	2016-03-31 11:56:45	2016-03-31 11:56:45	2015-07-01 00:00:00	3
32	2016-03-31 11:56:45	2016-03-31 11:56:45	2015-08-01 00:00:00	3
33	2016-03-31 11:56:45	2016-03-31 11:56:45	2015-09-01 00:00:00	3
34	2016-03-31 11:56:45	2016-03-31 11:56:45	2015-10-01 00:00:00	3
35	2016-03-31 11:56:45	2016-03-31 11:56:45	2015-11-01 00:00:00	3
36	2016-03-31 11:56:45	2016-03-31 11:56:45	2015-12-01 00:00:00	3
37	2016-03-31 11:56:45	2016-03-31 11:56:45	2015-01-01 00:00:00	4
38	2016-03-31 11:56:45	2016-03-31 11:56:45	2015-02-01 00:00:00	4
39	2016-03-31 11:56:45	2016-03-31 11:56:45	2015-03-01 00:00:00	4
40	2016-03-31 11:56:45	2016-03-31 11:56:45	2015-04-01 00:00:00	4

Fig. 5.10 Structure of the "eco_intro" table

Fig. 5.11 Graphic representation of the web portal

The last date, when information was collected, is displayed as standard. It can be changed the date with using the search form. Also in this form, it can be specify whether or not to display the structure of triangles created by the Delaunay triangulation method, display the values and color design of the calculated points and display a color map of pollution.

In the history of data entry, it can be viewed and edit the input data and the calculation of the degree of contamination in a specific PMP or NMP with a specific date.

5.3 Conclusions

1. The functioning of the state environmental monitoring system in Ukraine is currently unsatisfactory. The implementation of ecological information systems and software products based on modern geoinformation and web technologies for the prompt solution of monitoring tasks and informing the population about the state of pollution, including SLAA, is becoming urgent.
2. The purpose of creating an environmental monitoring system of SLAA is to organize, on the basis of geoinformation and web technologies, an intersectoral and hierarchical system for collecting, processing, storing and issuing information, which provides continuous diagnostics of the state of the environment and information support for decision-making aimed at ensuring the environmental safety of an urbanized area.
3. A necessary and sufficient condition for the functioning of the proposed integrated software solution for assessing the SLAA and its reflection in the GIS is a sequential solution of the stages, which include: ecological zoning of territories according to the I_{haa} excess indicator, interpretation of zones according to the I_{haa} excess indicator in environmental risk zones, and a mechanism for displaying environmental risk in a multilevel GIS matrix.

4. The scheme of a software solution for environmental monitoring and increasing the awareness of the population of UA, according to the state of the SLAA pollution, is presented on the basis of the Google Map software product and the application software interface. To ensure effective use of the software product, it is advisable to use a system of tables, which are interconnected by parameters.

References

1. Isaenko, V., Zaporozhets, A., Babikova, K., Gulevets, D., Savchenko, S.: Review of methods and means on monitoring the air pollution. Proc. Natl. Aviat. Univ. **3**(80), 61–70 (2019). https://doi.org/10.18372/2306-1472.80.14275
2. Zaporozhets, A., Babak, V., Isaienko, V., Babikova, K.: Analysis of the air pollution monitoring system in Ukraine. In: Babak V., Isaienko V., Zaporozhets, A. (eds.) Systems, Decision and Control in Energy I. Studies in Systems, Decision and Control, vol. 298, pp. 85–110. Springer, Cham (2020). https://doi.org/10.1007/978-3-030-48583-2_6
3. Kobus, D., Kostrzewa, J.: The use of spatial data processing tools for air quality assessments—practical examples. Inf. Syst. Manage. **4**(4), 251–263 (2015)
4. Morawska, L., Thai, P.K., Liu, X., et al.: Applications of low-cost sensing technologies for air quality monitoring and exposure assessment: How far have they gone? Environ. Int. **116**, 286–299 (2018). https://doi.org/10.1016/j.envint.2018.04.018
5. Arroyo, P., Lozano, J., Suarez, J.I.: Evolution of wireless sensor network for air quality measurements. Electronics **7**(12), 342 (2018). https://doi.org/10.3390/electronics7120342
6. Aleksandrova, T.N., Lipina, L.N., Grekhnev, N.I.: Geoecological estimate of state of nature in a mining and processing plant influence area using geoinformation technologies. J. Min. Sci. **49**, 167–174 (2013). https://doi.org/10.1134/S1062739149010206
7. Zou, B., Zeng, Y., Qiu, Y., Liu, H., Zhang, H., Hu, H.: Spatial and temporal variations of air pollution exposure risk based on geo-information technology Aided Proximity Model. In: 2010 18th International Conference on Geoinformatics, pp. 1–4. Beijing (2010). https://doi.org/10.1109/geoinformatics.2010.5568114
8. Merem, E., Robinson, B., Wesley, J.M., Yerramilli, S., Twumasi, Y.A.: Using GIS in ecological management: green assessment of the impacts of petroleum activities in the state of Texas. Int. J. Environ. Res. Public Health **7**(5), 2101–2130 (2010). https://doi.org/10.3390/ijerph7052101
9. Zaporozhets, A.O., Redko, O.O., Babak, V.P., Eremenko, V.S., Mokiychuk, V.M.: Method of indirect measurement of oxygen concentration in the air. Naukovyi Visnyk Natsionalnoho Hirnychoho Universytetu **5**, 105–114 (2018). https://doi.org/10.29202/nvngu/2018-5/14
10. Zaporozhets, A., Eremenko, V., Redko, O.: Metrological assessment of the indirect method of measuring the concentration of oxygen in the air. In: 2019 IEEE 8th International Conference on Advanced Optoelectronics and Lasers (CAOL), pp. 640–643. Sozopol, Bulgaria (2019). https://doi.org/10.1109/caol46282.2019.9019506
11. Bai, Y., Zeng, B., Li, C., Zhang, J.: An ensemble long short-term memory neural network for hourly PM2.5 concentration forecasting. Chemosphere **222**, 286–294 (2019). https://doi.org/10.1016/j.chemosphere.2019.01.121
12. Xiaojun, C., Xianpeng, L., Peng, X.: IOT-based air pollution monitoring and forecasting system. In: 2015 International Conference on Computer and Computational Sciences (ICCCS), pp. 257–260. Noida (2015). https://doi.org/10.1109/iccacs.2015.7361361
13. De Montis, A., Caschili, S., Mulas, M., Modica, G., Ganciu, A., Bardi, A., Ledda, A., Dessena, L., Laudari, L., Fichera, C.R.: Urban–rural ecological networks for landscape planning. Land Use Policy **50**, 312–327 (2016). https://doi.org/10.1016/j.landusepol.2015.10.004

14. Vimal, R., Mathevet, R., Thompson, J.D.: The changing landscape of ecological networks. J. Nat. Conserv. **20**(1), 49–55 (2012). https://doi.org/10.1016/j.jnc.2011.08.001
15. Watts, K., Eycott, A.E., Handley, P., Ray, D., Humphrey, J.W., Quine, C.P.: Targeting and evaluating biodiversity conservation action within fragmented landscapes: an approach based on generic focal species and least-cost networks. Landscape Ecol. **25**, 1305–1318 (2010). https://doi.org/10.1007/s10980-010-9507-9
16. Iatsyshyn, A., Artemchuk, V., Zaporozhets, A., Popov, O., Kovach, V.: Mathematical approaches for determining the level of impact of ash-slag dumps of energy facilities on the environment. In: Babak V., Isaienko V., Zaporozhets A. (eds.) Systems, Decision and Control in Energy I. Studies in Systems, Decision and Control, vol. 298, pp. 1–13. Springer, Cham (2020). https://doi.org/10.1007/978-3-030-48583-2_1
17. Iatsyshyn, A., Iatsyshyn, A., Artemchuk, V., Kameneva, I., Kovach, V., Popov, O.: Software tools for tasks of sustainable development of environmental problems: peculiarities of programming and implementation in the specialists' preparation. E3S Web Conf. **166**, 01001 (2020). https://doi.org/10.1051/e3sconf/202016601001
18. Fábos, J.G., Ryan, R.L.: International greenway planning. Landscape and Urban Plann. **68**(2–3) (2004)
19. Tmusic, F., Manfreda, S., Aasen, H., et al.: Current practices in UAS-based environmental monitoring. Remote Sens. **12**(6), 1001 (2020). https://doi.org/10.3390/rs12061001
20. Fraisl, D., Campbell, J., See, L., et al.: Mapping citizen science contributions to the UN sustainable development goals. Sustain. Sci. (2020). https://doi.org/10.1007/s11625-020-00833-7
21. Stewart-Oaten, A.: Chapter 2—Goals in environmental monitoring. In: Detecting Ecological Impacts. Concepts and Applications in Coastal Habitats, pp. 17–27 (1996). https://doi.org/10.1016/B978-012627255-0/50004-5
22. Babak V.P., Babak S.V., Myslovych M.V., Zaporozhets A.O., Zvaritch V.M.: Simulation and software for diagnostic systems. In: Diagnostic Systems For Energy Equipments. Studies in Systems, Decision and Control, vol. 281, pp. 71–90. Springer, Cham (2020). https://doi.org/10.1007/978-3-030-44443-3_3
23. Iatsyshyn, A., Popov, O., Artemchuk, V., Kovach, V., Kameneva, I.: The peculiarities of the thermal power engineering enterprise's ash dumps influence on the environment. Prob. Emerg. Situations **28**, 57–68 (2018). https://doi.org/10.5281/zenodo.2594489
24. Shkitsa, L.E., Yatsyshyn, T.M., Popov, A.A., Artemchuk, V.A.: The development of mathematical tools for ecological safe of atmosfere on the drilling well area. Neftyanoe khozyaystvo—Oil Industry **11**, 136–140 (2013)